SAILING MARMAX

A Spirited 218-Day Ocean Journey That Shares The Delights And Adventures Of A Kiwi Family Who Sailed Their Yacht From Spain Back Home To The Wild Waters Of New Zealand

DEBBIE ALDRED

ISBN: 979 8 2236 9363 5 - Paperback

Self Published by Debbie Aldred
PO Box 81, Matakana, Auckland, New Zealand p:
+64201 406 967   e: debbie.aldred@gmail.com
For further copies, Google **Sailing Marmax**

Dedication

This book is dedicated to Glenys, Bruce, and Janelle.

Every day, I pause to dip my fingers
into my magic bowl of tiny Kauehi shells.
I close my eyes, grateful that we have been truly blessed
to have all made it home safely.
Thank you for choosing Les and me to join you
to experience the delights and dangers of a true sailing adventure.
This book is my gift to you…

*"If somebody offers you an amazing opportunity,
but you are not sure you can do it,
say yes – then learn how to do it later!"*
-SIR RICHARD BRANSON

.

Contents

Prologue

I t really is funny how things work out…

During the Christmas holidays of January 2018 in New Zealand, my partner, Les, and I were cruising for ten days aboard our beloved 32-foot yacht, *Shadow*. Our sailing buddies were my sister, Glenys; her husband, Bruce; and young daughter, Janelle. They were aboard their beautiful 42-foot Moody design yacht, *Arctic Lady*. Most Christmases we were up in the famous Bay of Islands way up north. However, this year, it was Great Barrier Island, a beautiful and rugged island in the Hauraki Gulf of New Zealand, a well-loved playground for yachties from around the world.

It was early afternoon in picture-perfect Two Island Bay, otherwise known as Kiwiriki Bay. Les and I were relaxing in the cockpit of *Shadow*, doing the usual survey of the bay, whereby you get to leisurely check out who your new floating neighbours are. My eyes rested on the largest yacht in the little cove, a long steel beauty. Clearly, an overseas boat. She was rigged with all her offshore safety and navigational gear and was not a recognised New Zealand design. Curiously to me, a lady was sitting alone on the bow, gazing at the stunning scenery surrounding her. A man in the cockpit was playing around with what looked like a fishing rod. Now, I don't know if it is normal behaviour or not. I have this habit of dreaming up imaginary characters and wild stories about boats and people if I don't know anything about them, just for entertainment. It is a weird habit; my sister does it too.

From what country had this boat sailed? Who was aboard? And where were they heading? In my imagination, the couple, clearly not speaking to each other, had just endured a very long journey from America through the Pacific Ocean and down to New Zealand. Perhaps it was a rough trip? I couldn't imagine what it would be like to be on a yacht for months on end, having to endure stony conversations with a husband you didn't particularly like. I bet she was coerced into this ambitious man's wild ocean adventure; I bet she was over the trip and not speaking to him now. Perhaps they were lonely? Maybe they needed company? Maybe a drink?

Arctic Lady was anchored a few boat lengths from us, and Janelle was gliding past on the glassy water in her kayak.

"Hey, Janelle!" I yelled. Her grinning face came up alongside *Shadow*. "Hey, could you please do us a favour? See those lonely-looking people over on that boat there? Can you ask them if they want to come across for drinks aboard *Shadow* at five pm? And please ask your mum and dad to come over too."

Off she paddled up to the "American" boat, returning a thumbs up to us as she returned to *Arctic Lady*. I could not wait to hear all about their offshore adventures!

After a quick chat about the lonely ocean adventurers coming across for a drink, Bruce and Glenys settled into our tiny cockpit to partake in a beer or two. The American couple clambered into their dinghy and drew up alongside. "Welcome aboard!" we sang in unison.

Les extended his hand to welcome them aboard. (Les is Australian, by the way.) "Gidday, mate, I'm Les," he greeted them with a beaming grin.

"Craig and Carey, thanks for inviting us across!" My ears twitched; my eyes rolled skyward. Hmm…that was not an American accent!

"Where are you from, mate?"

"We're retired dairy farmers from Pukekohe," he replied in the broadest Kiwi accent on earth!

A stunned silence, then a gale of laughter followed. What the? Dairy farmers from Pukekohe? I could not believe it! My exotic Americans had morphed into Kiwis who lived only an hour from home! I explained my silly story to them. To cut a long story short, the evening was a total success. Much banter and hilarity followed, as it does in New Zealand when on holiday on a yacht and full of beverages!

We had just met Craig and Carey Staunton off their 52-foot yacht, *Maamalni*, which they had, incidentally, just brought up from Auckland. This was their first venture to the scenic islands of Great Barrier. New to blue water sailing, they were hoping to go cruising up the Pacific Islands in the future, which is why they had bought the boat. Well thought out, Debbie!

Little did I know that day how this chance meeting with this adventurous couple would change the path of our family's life over the next twelve months.

Come early June of 2018, Craig and Carey were planning their first trip to Tonga and needed a reliable crew. I had previously and eagerly offered my crew services to them. However, I could not make it due to work commitments, so my brother-in-law, Bruce, stepped up to the plate to help them out. They joined the Island Cruising Association (ICA) in the 2018 Pacific Rally, leaving from Opua, New

Zealand, and cruised up to Tonga with a fleet of thirty-four other yachts. While anchored in a stunning lagoon in Tonga, Bruce fell in love; he just didn't know it at the time, but let's say he developed an obsession. He spotted two 54-foot Oyster yachts, one called *Babe*, the other *Oyster Reach*. The seed was sown.

After the rally, Bruce flew home to New Zealand, and a series of life-changing events began to unfold. He had just turned sixty years of age on May 13. It was time for him to reflect on his past, plan for the future, and have a crack at some new goals. Bruce had held the position of Commodore of our local yacht club for some six years. He felt it was time for fresh minds to tackle the challenges of a swiftly growing club and the new marina development and had consequently just retired from the role. Bruce's mother now required full-time care. The harsh reality of observing the growing ill health of close relatives and friends was also a motivating factor. Glenys and Bruce's daughter, Janelle, now fourteen years of age, was not making headway at school, and it was worrying them. With all these things now playing on Bruce's mind, the thought of owning one of those shining Oyster yachts found bobbing in the sunny lagoon of Tonga began to look extremely attractive.

Searching the internet for options, in August 2018, Bruce found an Oyster 46 yacht. The problem was it was located in Spain…yes, on the other side of the world. How would he tell "the handbrake", my sister, Glenys, Bruce's beloved wife of fourteen years, he wanted this yacht? To escape the bitter cold winter in New Zealand, Bruce, Glenys and Janelle had just returned from a fortnight's holiday in sunny Rarotonga. Bruce had also just given himself a sixtieth birthday present in the form of a hot red Jaguar SUV. What about *Arctic Lady*? Glenys would not be amused!

A series of excited, covert conversations went on. Janelle knew about Bruce's new dream, Les and I also learned about it, and probably a few others too! Eventually, Glenys found out. In her own ever-so-formal words, "My allegiance to my already perfect lifestyle is at stake." Glenys's thoughts were: Why waste money on a new yacht when they already owned a beautiful boat? Glenys loved *Arctic Lady*. Summer in New Zealand was approaching fast. Our favourite cruising time was in the emerald waters of Kawau, Great Barrier Island, Bay of Islands and our magnificent Hauraki Gulf coastlines. Glenys was also fully committed to our local yacht club centre boarders. Teaching sailing, chief organiser, fundraiser, secretary, and do everything, the go-to person for our junior sailors aged eight to fourteen years. Who would take over her voluntary position with so much work left to do? Who would care enough? How could their little family of three get a 19-tonne, 46-foot yacht home from Spain, for goodness' sake? Pick up crew all the way? How long would it take? How would they find a team who could afford to take seven months away from work and their lives?

Les and I were already one step ahead of Glenys. We had already gone through the pros and cons of such an ambitious sailing adventure. The thought of bringing a yacht home from Spain back to Auckland excited the living daylights out of me!

As a family, we were all very aware of the financial implications, the actual personal risks and the dangers we would face. But we still sold Glenys the sizzle, and she eventually agreed to support Bruce's hair-brained idea to sail the Oyster from Spain to New Zealand in 2019.

All they had to do was purchase her, and her name was *Marmax*. How did we convince a very headstrong Glenys that it was okay, and we could do it?

1. So what? It is only money, Glenys. Sell another investment and live!

2. Sell *Arctic Lady*; you'll fall in love with the next yacht anyway.

3. There will be plenty more summers to enjoy on the ocean. The Caribbean and Pacific sound like a pretty good place to be during the coming New Zealand winter. Follow the sun, Glenys.

4. Having owned several businesses, we know that no matter how worthy one person is, no one is indispensable. After years of Glenys's yacht club volunteer work, it is time for a new person to take over anyway. Suck it up and see, hey!

5. Bruce's mum is in good hands, and our mum is in excellent health. Seven months will fly. (If our mothers had not been in such good health, we really could not have even contemplated this trip.)

6. The life experience for Janelle will be incredible! She'll be right!

7. Les and I fix their crewing problem. Glenys and I have always sailed together; Les and Bruce have gotten to be pretty good mates while racing and cruising (and drinking) aboard *Arctic Lady*. Apart from that, we are all family and know each other well enough to be able to handle the ups and downs of life at sea for months at a time. (Well, we certainly hoped we would!)

Besides all that, would Les and I let them go on their own? Hell no!!!

Bruce and Glenys jumped on a plane on October 26, 2018, and headed for L'Escala, Spain. They came home with a much lighter sack of gold and the biggest grin on their faces. They had put a deposit on the boat, and we were to welcome the good ship *Marmax* into our family.

"If somebody offers you an amazing opportunity, but you are not sure you can do it, say yes – then learn how to do it later!" The teasing words of Sir Richard Branson went over and over in my head.

When suddenly faced with the reality of the situation, oh wow... the "what ifs" started sinking in. Bluewater sailing around the world has some serious risks. Five members of our family were about to jump aboard a 46-foot piece of fibreglass. The ocean was enormous. What if we hit a shipping container and sunk? What would happen if pirates attacked us? What if we were dismasted in a wild storm? Would we make it home safely? All these thoughts rushed through our minds.

A few non-sailors thought we were pretty foolish, but most of our friends were so excited for us. This adventure was something many people had dreamed of doing but were too afraid to try. We walked into our local yacht club; I could feel the curious eyes upon us. Well-wishers hugged us, shook hands, and a few tears...clearly, some of these people never expected to see us again!

It was slightly weird how suddenly my world became more "sensitised." Les and I managed the family farm in Matakana, New Zealand. Walking up the driveway, kicking stones, it dawned on me that the grass seemed so much greener. Sounds silly, but as the deep eyes of our beautiful animals looked up at me in the yards, I wondered did they too know we were leaving? Why were the leaves on the trees so much softer to touch? The dirt, while pulling up potatoes, so rich, warm and loamy. The yacht club marina's water glowed gold on sunset. It was like, suddenly, we were leaving a beautiful planet and going off for a wild trip to another, leaving everything we loved behind. It was a feeling quite indescribable and weird; I can only imagine what it was like for astronauts leaving Earth. Confronting and very thought-provoking.

I worked in real estate. That was a fun conversation, telling the team we were very soon going to have a nine-month break and sail a yacht halfway around the world. My clients had to be informed. We had to find a farm manager. Storage for cars, empty the house for the new farm manager, organise house sitters, would they be great dog sitters as well? As vice-commodore of the Sandspit Yacht Club, I also had to find someone to represent me. Another fun conversation with the committee! You soon realise the implications of randomly dropping out of society when you have a busy life and a large family. Our blended family of six kids were fully supportive. Our seventy-eight-year-old mum, previously an adventurous sailor herself, was not exactly thrilled about us going. Still, she knew how exciting the trip would be. We had our younger sister, Helen, staying behind to support her. We were fortunate to be recommended a great farm manager to watch over both the farm and mum while we were gone, as long as we came back.

Mum pushed her point over and over again…"Everyone is horrified that you are going over to the Mediterranean in the middle of February. It's in winter. You are crazy!"

Well, Mum, you know we are crazy. I think we got it off you!

Come January 26, 2019, Glenys, Bruce and Janelle boarded another plane. Laden with stainless steel yacht fittings, minimal clothing, and boat gear, their destination? Barcelona, Spain. *Marmax* was sitting on a marina in L'Escala. This was some 100 nautical miles north of Barcelona, where they picked her up and sailed her down the frigid, icy cold and windy coast to Barcelona.

Les and I flew from Auckland, New Zealand, via Singapore to Barcelona on Thursday, February 14. I do have to tell you this story, even though it was rather embarrassing!

Les and I decided to commit to this exciting sailing adventure, so I bought the air tickets...online. I'm a busy person. Between business calls, I organised two airline tickets. Mistake 1: The autofill popped in Debbie Lee Aldred, instead of my real name, Deborah Lee Aldred. Did I notice? No. I was totally oblivious to the rigid rules of international travel!

A second ticket was ordered for another. Autofill, Les John Marsh. Mistake 2: it should have been Leslie John Marsh. I pushed the go button on my Visa card for two thousand dollars. Hold on, let me answer another call. I printed out the tickets. We were all set to fly to Barcelona. Were we excited? Yes! Hold on, our names had to match our passports. Could I correct our names on that printout? Of course not. I phoned, I emailed, I faxed the travel company for four months!

Adopting the popular, "she'll be right" attitude, Les and I fronted up for check-in at Auckland International Airport, 6 am on the big day. This was the start of our epic adventure. Long story short, we bought two crispy new air tickets for another couple of thousand NZ dollars. The real Deborah and Leslie managed to land in Barcelona at the exact arrival time, as planned. Lesson learnt the hard way!

MARMAX'S LINES

LENGTH OVERALL: 14.26m - (46' 10")

BEAM: 4.41m - (14' 6")

DRAFT: 2.16m - (7'1")

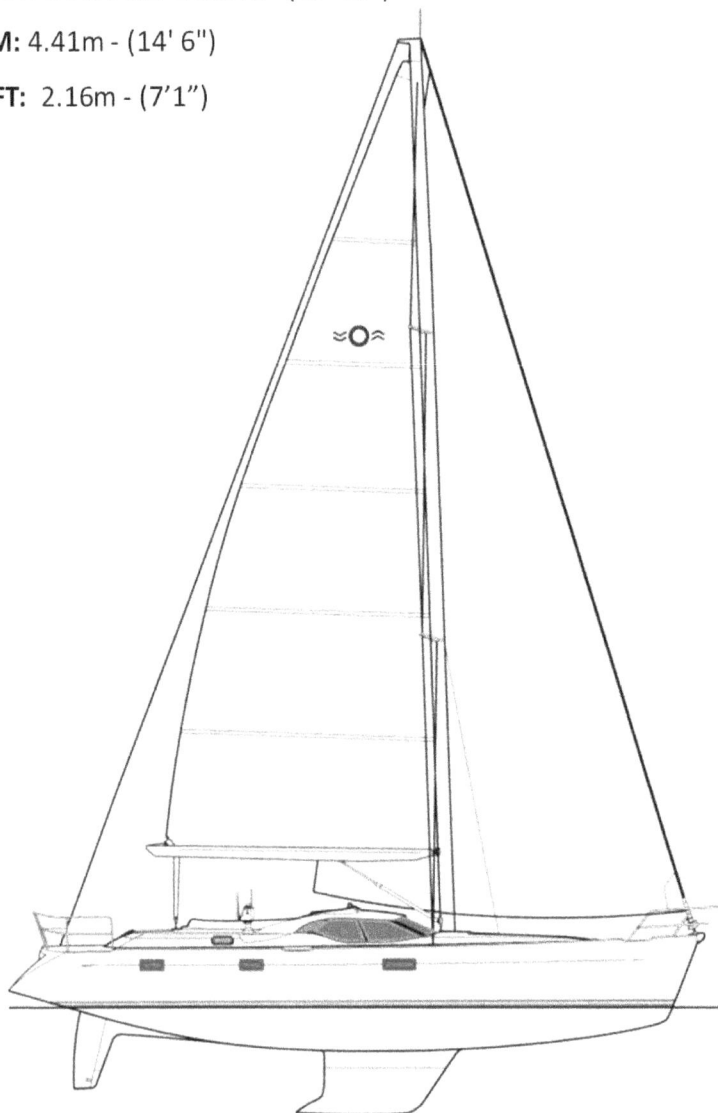

MARMAX'S INTERIORS

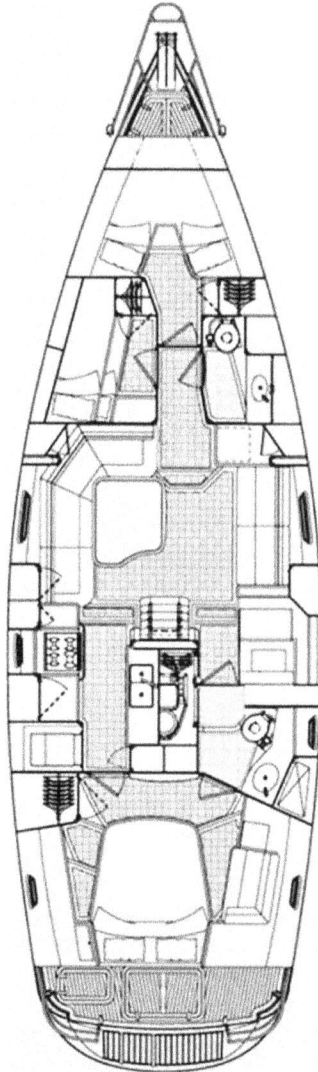

1. The Marmax Story Begins

We arrived in Barcelona to meet the beautiful yacht *Marmax* and reunite with family: Glenys, Bruce, Janelle, my sons Sam and Nick, and their partners, Emily and Tash. Just thought I'd introduce them that way, as you will start wondering how we all fit in!

Two of my sons and their partners worked on the superyachts based in Europe, so how lucky were we that one of our son's vessels was moored on the very same marina as us. It had long been a dream of ours to catch up with these guys. They had all been away from home for a couple of years. The crews put on a fabulous welcome BBQ for us with great food and good old Aussie and Kiwi rock music in the background. It was amusing to hear this type of music in a land where no one seemed to speak our language. Les and I were not experienced world travellers by any stretch of the imagination.

The Port Fòrum marina was glorious and only a couple of minutes' walk to everything we needed. The metro transport system was also only five minutes away. It was chilly! Four degrees at night to fourteen degrees during the day, sunny and delightful! We had plenty of clothes, so as long as we dressed appropriately, the days were superb! Time to pay the final boat bills, buy a Spanish flag and a blow-up kayak for Janelle.

Marmax had just come down from being on the hardstand in the boatyard. Her big mast was to be re-rigged for New Zealand category 1 insurance. She was floating high on her waterlines with no fuel, water, or mast in place, kind of lonely looking. We topped her up with diesel and water and did a thousand other jobs, including endless shopping trips for food and storage containers for it all. Glenys and I stored the food into zones from one to six. This system would save us time and also prevent us from foraging endlessly for supplies and upending cans, packets and jars in violent seas once we got going. We stuck the long hand-scribbled inventory list to the galley (kitchen) wall.

After the mast popped back in, we had sea trials from the Port Fòrum marina. This ensured that all the rigging, sails, winches and lines were working in perfect order. Oops, a rookie error: we had forgotten to fly our national flag off our stern, which would identify us as a New Zealand yacht. Instead, we had our much-revered, black and white silver fern flag hoisted up, which appeared to be unidentifiable to the Guardia Civil (Coastguard). They came screaming alongside us in their boat at sea, barking out a stern order to get immediately back into port for boarding. Thank goodness our English/Spanish-speaking riggers managed to defuse, potentially, a dire legal situation for us. Unrelenting eyes scrutinised our precious passports, and we were thankfully released. Let's get out of here! We proudly raised our New Zealand national flag and breathed a massive sigh of relief!

2. Marmax Arrives in Mallorca

We finally left the Port Fòrum marina in Barcelona on Tuesday, February 26. We were not entirely on time as, Barcelona worked on a time frame approximately half of the speed at which we operated in New Zealand. It took time to adapt to this work ethos, but one simply had to go with the flow!

On the positive side of things, we got to spend an extra four days with our kids. They took us to the most glorious places, including a quick flight to Tuscany, Italy, to catch up with Sam and Emily. Up into hills, gorgeous little restaurants, ancient towns, massive parks with every curiosity in architecture, food and people you can imagine: amazing!

After a farewell lamb roast dinner with the kids and a tearful goodbye, the blaring sound of ship horns serenaded us out of the harbour; we were on our way!

Blessed with the very best winter weather we could ever hope for, the Mediterranean Sea was a shining replica of tourism postcards, the sky a bright, magnificent blue. We began our gentle 100 nautical mile journey across to the island of Mallorca. Leaving at 1100 hours, we had motor-sailed through the night under the light of the evening star; the skies were truly stunning. There was no moon but just beautiful. We dutifully carried out our helm watches but overlapped each other's time slots as we were too excited to sleep. None of us wanted to miss the sunrise or the magnificent view as Mallorca came into our sight in the wee hours of the morning. God, it was cold. We passed several boats, mainly the big, fast-moving inter-island ferries.

How do you begin to explain the beauty of this place? The mountainous windswept cliffs were impressive, to say the least, with deep drop-offs and no reefs, rocks or navigation hazards to worry us at all. We were on our way to the main port of Palma to pick up a new life raft that had not reached us on time for our departure in Barcelona. Along the way, with very little wind at all, we motored close to the stunning coast to check out the many quaint ports and fascinating historical towns. The mansions dripped over the edge of the imposing cliffs, making us feel very, very small. The superyachts were, well…there were no words for them. The wealth was pretty mind-boggling, and it did make one wonder, what was it all for?

Two standout ports were Port d'Andratx and Port Adriano in El Toro. Hand on heart: Stunning!

They were just two of the superyacht havens on this beautiful island. It looked like a few houses may have slipped into the ocean recently with lots of new builds going on. Money, money, money. Because it was winter, most of the mansions were empty; our voyage was most timely! We could only imagine how busy it would be when the European tourists arrived in the summer. It was fabulous to cruise around these lovely coastal places on our own; it certainly was another world from where we came.

Phoning ahead for a marina berth, we slid through the harbour entrance into the impressive port of Palma, the largest town of Mallorca. The first quote for a one-night stay was €130. Wow! That was NZ$216 per night! Glenys got on the VHF radio. We scored another berth right outside the yacht club, the Reial Club Nàutic de Palma, for €60 per night, which was still NZ$100 per night, but this was Mallorca! While sailing, we obviously had to keep an eye on our budget. We were not earning an income for the entire year, and the cost of travelling through countries could often be high and full of unplanned expenses.

The marina manager asked us to park in backwards, between two rather large, expensive-looking yachts parked on the marina. How on earth would we fit in there without damaging not only our hull but other parallel boats? Bruce's fancy bit of reverse throttle and bow thruster work helped, with a bit of nudge from a fearless young man in a high-powered manoeuvre tender. We had landed. It was a bit like ramming a stubborn cork back into a wine bottle! Our newfound English friends berthed next to *Marmax* informed us that in the height of summer, €600 per night was the norm for them and their 54-foot yacht. What a set-up this yacht club was. So classy, so historic, welcoming, full of memorabilia and charm; best of all, they had excellent wi-fi! I had just started a blog about our ocean adventure. I was now under my self-imposed pressure to get it out, on time, to our eager families, friends and many followers.

We spent the entire day walking around this jaw-droppingly gorgeous city. If one was to ask which was the more beautiful, Barcelona or Palma, I'd say equally stunning! Janelle was sick of seeing castles. She had been visiting them worldwide all her short and privileged fourteen-year-old life. Les and I, yep, the virgin European tourists, walked around with our jaws on the ground. If I had said the word amazing once more, I would have had to consider hitting my head with a winch handle. There were no other words for it, and to see it all under such clear blue sunny skies was such a treat. The twelfth-century gothic Palma cathedral, the tenth-century Arab baths, monasteries, wonderous historic streets with stunning architecture, local markets were bursting with the most incredible and colourful delights!

The Mallorcans seem to have a much sweeter tooth than the Barcelonians; lollies, delicious pastries, nuts and goodies gleamed longingly at us. They love the fragrant smell of candles and soaps too. We discovered the most oversized doughnuts we had ever seen, loaded with gooey, decadent, rich chocolate. Mmmmm. The golden olive oils on freshly baked street-oven bread…do not get me started.

On the morning we were hoping to leave, we woke up to the sound of fog horns going off everywhere and a total whiteout; we could not even see our bow, it was so dense. Bruce just had to go to the marina office to book us for another night; it had lifted by early afternoon, but what a curious sight it was.

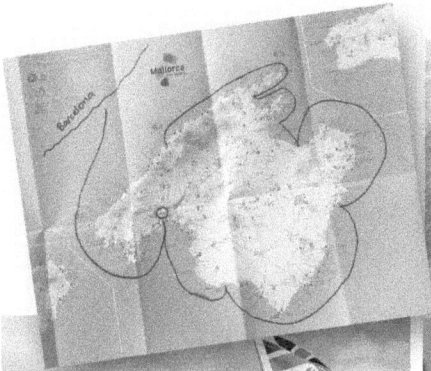

Port D'Andratx

12th Century Palma Cathedral

Always watching for rocks!

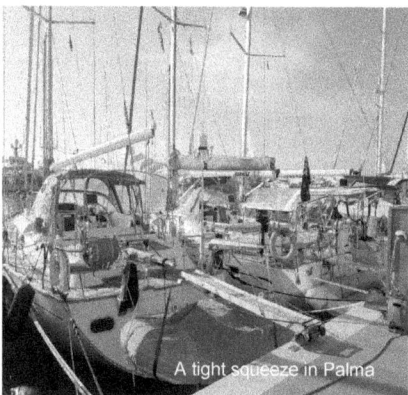
A tight squeeze in Palma

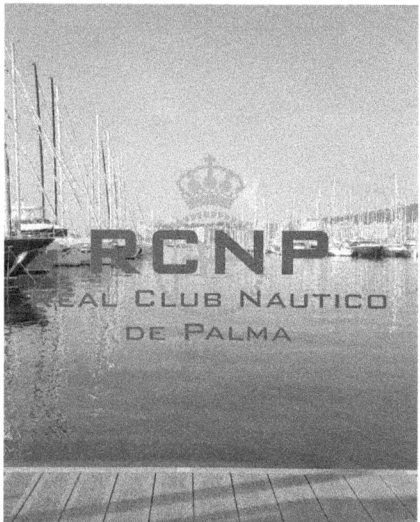
RCNP
REAL CLUB NAUTICO DE PALMA

3. Heading North to the Port of Pollenca

After a quick trip for fresh food supplies and farewell coffees with our newfound friends, we finally departed on the next leg around the island of Mallorca.

A steady breeze had us cruising nicely around the rugged coast heading southeast. Deep blue waters in close, and no rocks or reefs, allowed us to drift directly under the ledges above. All the way, hotels and buildings hugged that coast, many under construction with giant cranes swaying above them, large and small, with cute, white homes clinging to the precipices. Homes had been constructed precariously close to the edge of the cliffs, others clinging to the vertical. Why? Not sure. There was lots of other dirt to build on and, really, only a lonely ocean to gaze upon if you lived there. You did have to shake your head and wonder.

We quickly realised that the rave reviews of some very appealing-sounding anchorages were focused on much smaller boats than

Marmax. In fact, they were more suited to our inflatable tender. After several attempts at motoring up to several spots, we found that we'd be backing out of a squeezy space if we entered. We soon learned to look at the photos in guidebooks from a different perspective. It cut our snooping time around by half. It was so exciting sliding into these little rocky anchorages and not knowing what we would find; many just took our breath away. Every time we gushed, "This is the most beautiful anchorage in the world," another beat it around the corner. Yep…like I said, truly amazing!

Mallorca is 1400 square miles in size. The four-day cruising itinerary suggested in the guidebooks for circumnavigating the island was once again designed for speed boats, not sailors. Similarly, the anchorages they raved about were mostly for little fishing boats and small pleasure boats, not for us. The trip took a little longer than we expected because we wanted to see all the hidden spots, big and small, enjoy the ports as much as possible, and sail whenever possible.

We did not always want to berth in a fancy marina. First up, they were costly, but there were very few sheltered spots around the island and even fewer places where you were allowed to anchor. Resorts seemed to own the space out the front of their establishments. You could not drop an anchor where there were native seagrass beds, and there was no fishing anywhere without a permit. We decided it was calm enough to anchor off this long sandy beach on our first night out of Palma.

There were no resorts in sight, but there was another large yacht anchored closer to the shore reassuringly. We checked to make sure we had not dropped our anchor amongst the seagrass, put on our night lights so we would not be mowed down by another boat in the dark, and settled in for the evening. The sun had set, the night was closing in, and we noticed this little timber fishing boat chug-chugging out from the shore towards us. No navigation lights on, going ever so slowly and stealthily towards us. It approached us from behind, came right alongside, and had a good nosey at our yacht. Our hearts were galloping, the air electric with tension. We were just a bunch of country bumpkins from New Zealand, not aware of whether or not there were pirates in these waters. Why were we attracting this unwanted scrutiny from this unidentified vessel? It started circling us, and by this time, we had decided to take action and stand guard as a united force in the cockpit and stare back at them. Two of us, armed with torches, began checking the hull to make sure these intruders were not planning an attack. Without a word, they chugged quietly off into the distance. We were unsure what was happening. Were we in trouble and having our identity reported to the authorities for anchoring near the protected seagrass? Were they planning an attack, or were they innocently curious about our yacht flying the New Zealand flag? We never stuck around to find out. We were gone by daybreak, so the Guardia fellas could not catch up with us again. Like a fox on the run!

As we were approaching the next little inlet, Porto Petro, a steady string of twenty-one yachts, in full sail regalia, popped from behind a rocky harbour entrance in readiness for a race, a magnificent sight indeed! Upon entering this curious harbour, we noticed there was not one yacht left in the marina. Still, there were probably close to a hundred small traditional fishing boats. It seemed, in this little port, it must have been compulsory to race. The water in this harbour was crystal clear. We had zigzagged the yacht through a labyrinth of submerged rocks and had ended up in a tiny, blue, crystal clear lagoon. Les volunteered to stay aboard while the rest of the crew excitedly headed for the shore.

Once there, we found a small trail that led us to an extraordinarily pretty little fishing village surrounding a small harbour. There were a few small shops to explore, a sleepy cafe and old Spanish villas clothed in colourful bougainvillaea flowers everywhere. Glenys suggested mischievously to Janelle she should take note of this place and come back here for her honeymoon. I cannot tell you how beautiful it was; you will have to take our word for it.

Up the coast, more amazingness. Several hours later, in the late afternoon, we entered a picturesque little village port called Porto Cristo. Equally stunning, but busier and more significant than Porto Petro. We reversed into the marina berth; there were no floating pontoons alongside boats here. Bruce was becoming pretty good at landing by now. Well, he had no choice, did he! We tied up aft then shimmied along the yacht, dragging up the bowlines from the seafloor, which ran the entire length of *Marmax*. These were often slimy with mud and barnacles, perhaps lacking use at this time of year, and always very heavy thick ropes, but we managed.

Then we had to erect our passerrail (gangplank) off the aft end, enabling us to step ashore. We carried our fenders (all ten of them) in our inflatable tender, which hung off the davits (swinging arms) off the stern of the boat. Most of this gear was packed neatly into a large, underdeck storage locker to the ships stern; this is called a lazarette. On an ocean passage, the lazarette not only protected everything from the weather, but it also gave us more space on the back deck. It also hid our treasures from the roving eyes of possible pirates, who, we were informed, often only ever want an outboard engine.

We had barely tied up in Porto Cristo, and Janelle was off to christen her new inflatable kayak. She was in it and gone. What fun! It proved to be a handy and crucial piece of equipment for the duration of our sailing adventure, especially for a fourteen-4-year-old girl travelling so closely with her family. Janelle is a quiet, shy, but adventurous girl. The only child of Glenys and Bruce, an experienced world traveller by land and a sailor for all her short but exciting life. She is very confident in her skin, does not need the security of friends to empower her and finds it very easy to keep herself occupied. This trait proved to be a tremendous asset to all of us, as we were to have many monotonous days of open ocean sailing between countries. She chucked only one teenage meltdown for the entire trip home; how good was that!

A quick tidy-up and change, and we were off to check out the town. Clean, tidy and welcoming. Plenty of lively cafes, bars and restaurants lining the stunning waterway. Sangrias, salmon and hamburgers topped off an eventful, fun day.

Back out to sea, we sailed along the raw and pristine coast for miles and miles. There was so much to see. Small beachside towns, full of pretty white houses. Sprawling multi-storied buildings to the water's edge, more cliff-hanging communities and sheer rock faces. Lighthouses, walled castles, monasteries in the mountains, and caves galore. You name it, we saw it. At one point, a fine mist of hazy green pollen drifted through the hillside gullies. We had sailed around the top of the island with a brisk offshore breeze, but once we reached port, the dust had settled on us, covering the boat with light green pepper.

We enjoyed idyllic sea conditions along the east coast, a gentle and steady breeze all day long. *Marmax* had her own Spotify music channel. We sang at the top of our lungs and howled with laughter; this was our kind of sailing, and we were finally starting to relax and get into the groove of life on a sailing yacht.

Without sounding idealistic, in reality, just two weeks before, we were all different people. I, for one, worked seven days a week. We were extremely busy people who packed so much into life; that's the way we liked it. We had stepped off a plane from the green grass of home, left all our creature comforts and changed our environment within twenty-four hours to live on a small yacht. It took me two entire days to stop looking at my phone, emails and messages, stressing because the communications were diminishing. We left Barcelona, and six hours into the trip, our phones stopped working. No more communication…zip. I had to learn very quickly to let go of my control and go with the flow. Adapt to the situation, be flexible.

As we were about to discover, the ability to adapt and become flexible became crucial. Our well-thought-out plans often had to change on the yacht as our route was always dependent on weather. Sometimes equipment broke down; there was no one in the middle of an ocean to fix things! No fridge? No gas for cooking? No problems. Adapt.

Many shops we were to forage through in the coming months often did not have what we required. Sometimes, we were not interacting with anyone on the "outside" for weeks on end. We did things for each other. Pretend for a moment you are on a boat for ten days with a massive storm outside. How would you cope? How would you find a way to adapt to the situation?

The other skill we learnt was the ability to slow down. Early in the trip, we were coming down from a massive wave of energy resulting from action-packed preparations to get us to sea. Care was taken with every movement we made aboard the boat; this also required slowing down. Attention had to be paid not only for our own safety but also for that of the crew.

The breathtaking beauty we experienced as soon as we sailed into Mallorca compelled us to look around and appreciate all that we had and all that surrounded us. We were all created with two ears and one mouth; listening with those two ears became crucial. We had to discover a rhythm of our own, tune into and communicate with each other. Even though we were all family, things were not always rosy. We were all at different stages of developing these crucial skills for the trip. You didn't have to like a bad situation, but you sure did have to go through it. There is no hiding in a small home surrounded by sea. We learnt to set our own pace, blend in with each other and go with the flow when things got taken out of control. Living on land, finding your rhythm is equally important. At sea, we never felt isolated because we worked all this out early. When we came into an anchorage, we would have to step outside of our comfort zone and knock on a stranger's hull to find a friend. We did not know how we would be greeted, or even if they understood English! There were no work breaks around a coffee machine at sea. Practise flexibility and slowing down in your own life and notice the changes.

The next stop for *Marmax* was the peaceful township of Sant Pere. What an intriguing place! The marina development here was sensational, but where on earth was everyone? In their defence, we were not there during the tourist season, but wow, it barely had a heartbeat. As it turned out, we were the largest boat in town. After tying up, we went into the village to check it out; it looked kind of sleepy. It was like those Mexican towns where you expect a ball of dried-up old prairie grass to blow down the deserted street followed by a pe-owww from some lonesome gun-slinging cowboy. Not a lot was happening here. Some nice-looking homes, but most were empty and in need of a bit of attention.

A few inhabited houses were tidy, with determined gardeners making the most of their not-so-fertile-looking back gardens. Wafting freesias from one beautiful garden had us following our nose to another. So different from any other town we had seen. We came across a small supermarket, and guess what? Among other delights, they sold rum and fish sauce, just what we needed! There were loads of seriously damaged boats in this port – all shapes and sizes. It looked like they had been battered against rocks and smashed beyond saving. A sorry sight indeed. It looked like it would get a bit blowy in these parts; we stayed just one night, and then we were out of there!

We enjoyed an excellent, swift sail up to the Port de Pollença, dodging tiny fishing boats all the way. This next destination was to be truly stunning!

Soller

Porto Cristo

Puerto Petro

Beautiful Pollenca

4. Leaving the Balearic Islands for Gibraltar

We ended up spending two nights in Port de Pollença. The marina manager gave us a berth at one of the big concrete fuel jetties, miles from the action! We were, however, always happy to do kilometres of walking when on land. I had re-commissioned my Fitbit to calculate steps and mileage; the numbers were getting, well, pretty impressive!

Most of the beauty of the Port de Pollença was all about the actual marina. Hundreds of fascinating boats, all shapes and sizes, from all around the world. There were no superyachts for a change but plenty of wannabe, little (expensive) superyacht-type launches. Many were from Germany and England, some real beauties.

The marina was divided into travelling yachts, permanent liveaboards, runabouts and the traditional fishing boats of Mallorca, the llaüt. The main esplanade was wide and geared up for roving tourist crowds in the peak summer season. Behind the tourism façade, the place was undeveloped and not near as clean and tidy as previous towns we had visited.

Les and I found the local markets ten minutes before they shut. Glenys, Bruce and Janelle had found us fossicking amongst leather handbags, giant dates and strawberries. They hustled us, unexpectedly, down the road, as only Glenys can do. In five minutes, we were on a really flash bus, heading for the main town of Pollença, seven kilometres away. The town centre was a wonderful, exciting lesson in medieval architecture and history, a culture beautifully blended over the centuries.

The highlight was walking up to the top of the hill via the well-worn 365 marble steps, called the Calvari Steps to the famous Calvary Chapel. We had previously seen this masterpiece in a shiny tourist brochure. When my sister sets her sights on a mountain, the family are usually forced to climb it; she has had this effect on us all her life. Huge, wide stone steps smoothed by millions of sandaled feet over the years stretched up the hill. These were fringed by traditional Mediterranean villas decorated in wild bursts of bright red, pink and purple bougainvillaea. The arduous trek up those 365 steps was a lot of fun, but ten times more fun to run down!

It was my bright idea to have lunch away from the town centre. "Let's eat like the locals!" We single-filed into this tiny sparse restaurant with no menu (because they don't) and a lonely waiter/bartender who knew no English. However, everyone knows the universal word "beer". So down we sat at our little table with matching pygmy chairs, and the mute waiter served us the usual tapas and cold beers. We were then presented with three more intriguing dishes, more beer and a bottle of rich red wine for the boys. We ate until the delicious food stopped coming; we could barely get out of those little chairs! Obviously, as those of you who travel will know, the midday meal was the main event of the day there. For a mere €7 each, we felt as if we had just had Christmas dinner. Quite fun, but we did miss our bus back to our boat. We all crammed into a little taxi with a nice man, and thirty minutes later, we made it back to *Marmax* feeling fat and happy.

The following day, we enjoyed yet another crystal clear, glassed out day on the sea as we made our way around to the delightful Port de Sóller. This place was magical! The marina was a lot smaller than Pollença, with more local boats than visitors' vessels. There were no facilities for yachties, as the authorities were in the process of redeveloping the harbour; we always looked forward to hot showers when we came into ports. The first decent-size fishing trawlers we had seen shared our jetty; vessels were painted all colours of a rainbow. I found an English-speaking crew member who was happy to practice his newfound language with me. These guys appeared to mainly trawl for sardines and for what the Aussies call coral prawns, a by-product in our countries. Tiny, tiny crustacean creatures by comparison with Aussie and New Zealand seafood.

Having previously designed trawler nets in Australia, I was aching to show the local deckhands how to mend nets and do things more efficiently. Their methods were age-old and ineffective by our modern standards. I would have loved to have had the opportunity to design some decent fishing nets for them too. But I guess at least they would never over-fish the Balearic Island seas. Although, in saying that, as much as it was beautiful in the Mediterranean, the sea life was almost non-existent, like a very clean aquarium before you put your fish in. I left the fishing crew feeling quite sorry for them, but they were happy with their toothless grins, and you must respect their traditions.

It was starting to warm up a little by now, with a few scattered tourists bravely playing in the icy waters near the shore. The Port de Sóller was a pristine beach promenade edged with an ancient esplanade, boasting more stunning old architecture. Staring about in fascination, we ambled through the rustic street markets.

Come morning, we leapt excitedly onto the little historic train that followed the shoreline and went for a clickety-clack, rollicking ride into the charming village of Sóller. This place was considered to be the most beautiful city in all of Mallorca. It was set in a lush, fertile mosaic valley overlooking the Mediterranean with the impressive Tramuntana Mountains shining as a glorious, gilded backdrop. The slippery, burnished paving stones got me thinking. When and if it ever rained in the town, walking in high heels would have been pretty terrifying, especially going downhill around here. Steeped with history, the steps were so smooth and cool to touch, you wanted to lay your cheek on them. Magnificent, ethereal structures wrapped their arms lovingly around the pretty township. Intricate detailing included my favourites, the delicate brass door knockers, the shape of fine ladies' hands. We had never seen so many orange, lemon and olive trees; it was like skipping around inside a child's colourful storybook. I read that the name Sóller was derived from the Arabic word "sulyar", which means the Golden Valley. That worked it was a real gem. Did I mention the pastries in this town? To die for! Once again, we ate like a herd of little piggies.

There were lots of classy clothes shops with beautiful displays in those gorgeous old buildings. Lots of artisans, churches and narrow, little cobbled streets.

Before heading back to *Marmax*, we spent the last hour picking up sea glass off the beach. Over the years, carelessly discarded bottles, jars and glasses had been tumbled by waves to form naturally frosted, colourful beach gems. Sea glass takes twenty to one hundred years to acquire its unique texture and shape. It is getting harder and harder to find as the world embraces recycling. Curious bystanders watching us must have thought us to be quite silly. It was kind of neat to imagine all the thousands of people over the years who had raised their lips to this pretty glass. We packed our glass treasures up into little jars as precious souvenirs of, what we thought, was one of the loveliest places we would ever visit in our lives.

At this stage, our LPG gas on board the boat was getting low. We had plenty of gas bottles, but they were all Spanish, and we had English gas accessories. *Marmax* had been built in England. Another rookie error on our part. Who has ever heard of gas bottles and their fittings speaking different languages? Not us! The international hunt was on for a multinational gas fitting, and we looked everywhere!

At 3 pm we cast off our lines again for the next leg of our adventure. Unbelievably, there was still no wind for the 100 nautical miles we had to cover to reach Ibiza, to the south of Mallorca. Another bright, calm and starry night. I came on watch at midnight, taking over from Glenys, who then stayed and chatted for half an hour. At 12.15 am, we completely lost our radar reading for no reason at all. Weird. We tried everything to get it working again, as rather large transporter ships were moving about at a rate of knots. We were always on the lookout for these vessels in case they ran us over. As it turned out, upon investigation, we had just entered the famous Triangle of Silence, el Triángulo del Silencio. A spooky phenomenon, a little bit like the Bermuda Triangle. It was well known that radars and compasses on ships and aeroplanes often went haywire without any scientific explanation in this location. My late father would have been very excited; our family is wild about UFOs and weird stuff.

Bruce wanted to be clear of Ibiza by nightfall to make up time, so we continued to an island to the south called Formentera. This island boasts some of the most beautiful beaches in the world. We dropped the anchor in a pure white sandy-bottomed bay. Because I cannot let my little sister beat me at anything, I joined her and Janelle for the most bone-chilling, icy swim I ever had in my life. The water was like freezing blue champagne. Why did we do it? Just so we could now say, "Yep, we've swum in the Mediterranean." Geez, the things we do!

Our next planned stop was back to the southern Spanish coast. Cartagena, here we come! Another twenty-four hours of motoring followed by six hours of happy sailing; yachties did not call this sea the Motor-terranean for nothing. By this stage, we had barely sailed at all throughout our travels. Les kept us entertained by catching the first fish of our trip on a spinner line right at sunset: a big, fat, juicy yellow-fin tuna. We also had the privilege of meeting some very friendly dolphins who kept turning up to stare at us in the sparkling, clear water. I think they enjoyed our squealing too!

As we cut through the calm waters into the port of Cartagena, we were confronted with the first fierce force of military we had come across. Not only was this a port city but also a massive naval base. We counted sixteen navy ships and two enormous superyachts, but the marina was occupied mainly by international pleasure yachts and catamarans. While backing into the marina, the wind blew up to 17 knots; it would have been rather pleasant to sail in.

We were greeted with a friendly welcome from the marina manager and the keen attention from liveaboards surrounding us, mainly Americans and Germans. Smiles dangled off their lips, clearly hoping they could share company and beverages with us. We were the new social "meat", a craved change for some of these seasoned vagabonds who had been living on this same marina for years. Arriving yachtsmen with tales of adventures afar obviously whet their hungry appetites for entertainment. We had arrived late in the day, around 5 pm, and were a little stuck for time. We gulped a quick sneaky beer in the saloon of *Marmax*. Then we hurried off for a speed-date with the ancient city of Cartagena, a town dating back to around 220 B.C. We were on a mission to get to the Caribbean now. We simply did not have time for socialising that night.

Les and I had never seen Roman ruins before, and this town was packed to the brim with them, all in various stages of discovery and restoration. We ambled among the ruins as the sun was setting in a golden glow over the city. As dusk fell, spears of light shone up from the rubble, revealing the romance of the pink travertine columns and vaulted galleries from centuries past, just magic. Cartagena did not enjoy the wealth of Mallorca. Still, to their credit, every street had big renovation projects going on to protect not only the ruins but the stunning facades of their intricate, historic architecture. As darkness approached, one could not help but notice the step up in street security; surly police and intimidating uniformed guards were everywhere. Our marina manager had sternly warned us about the bad parts of the town. His words of heed had stuck in our heads as night revealed a blanket of suspicious-looking young men in the streets. They clearly had their social troubles here.

Once we had had an eyeful of history and culture and the back streets got simply too spooky, it was time for a drink at a local bar. We loved the compulsory tapas, with a beer or wine idea; there was so much colour, music, food and Spanish frivolity to enjoy in this cosy establishment. That was our speed-date. Finito.

Early the following day, we bid farewell to Cartagena and thundered straight into a 20–25 knot wind. With the wind behind us, we were cracking along! Sheer cliffs lined with turrets and intimidating black cannons pointed down on us. This was the breathtaking Spanish coastline called Costa Tropical in Andalusia, a province of Granada. Strewn along the coast, we found more ancient cities with grand tourist accommodations fringing white sandy beaches. Highly intensive and terraced horticultural areas sat below the snowy Sierra de Lújar. The curse of the discarded plastic from this massive food-growing location was undoubtedly an environmental concern. State-of-the-art wind turbines proudly stood out amongst the relics of a golden past, and we were, unbelievably, being escorted by a pod of dolphins up on the bow. We could not paint a prettier picture on our forty-hour long ocean route to Málaga; who would imagine!

We tried to enter the port of Málaga. It appeared *Marmax* was too large for them to accommodate us in their marina. The wind had dropped right off again at dawn, so we ended up motoring into yet another "curious" port called Puerto de Motril.

We sauntered down the road to the central police station. A seriously sexy-looking policeman, in broken English, informed us with a gleam of deviltry in his eyes that we had no need to get Spanish customs clearance. He advised this could be completed in the Canaries. We had intended to clear customs before leaving Spain, hence our perceived need to get into one of the last ports on the Spanish coast. Our following country would be Gibraltar, which, at the time, was still part of the European Union. As we were all Kiwis and Australians aboard, we were authorised to enter without clearance from Spain.

We went for a power walk around the town; there was nothing to report here. A bland concrete jungle with a rocky, unwelcoming waterfront. Why anyone would live here was beyond our comprehension. Everything was shut up because it was not the tourist season here, but it was still ugly. The marina and yacht club, however, was pretty and very accommodating. Glenys and I agreed the best thing about Motril was that tall, dark and handsome policeman. We gave Malaga a miss. Let's soldier on to Gibraltar!

Gibraltar

5. Apes, Bananas, and a Roller Coaster Ride Down the Moroccan Coast!

Gibraltar, so much more than we would have ever imagined!

After a howling ride from Malaga, smack on the nose of a 27-knot westerly wind, we arrived with the boat still intact and gazed up in awe at that immense Rock of Gibraltar. A Swedish yacht fell in behind us as we dropped the mainsail and tried to find the marina Glenys had previously booked a berth in.

For those who don't know, Gibraltar is essentially a British military base, with four to five hundred troops stationed there at any given time. We had learnt that anti-submarine nets were put up across the marina entrance at night for security, so we slowed the boat down and cautiously moved towards the port. A colossal, giant black submarine was stationed at the first entrance wall, behind the military base net that protected that part of the harbour. Finding that marina entrance was another thing; it was disguised very well!

Masses of high-rise apartments, British garrisons, and lengthy old stone seawalls sat beneath the Rock. We could see yacht masts behind high-rise apartments. How on earth could we get in there?

The Queensway Marina was yet another jaw-dropping stunner, much like a mini version of our Viaduct Harbour in Auckland. The Rock of Gibraltar stood majestically behind it, and everything else quietly guarded it.

Whenever we came into a marina, all crew were on deck, preparing multiple fenders and ropes. It was so hard not to keep sweeping your head about to take in all the jewels we were presented with, gazing at the marvellous marinas and their historic surroundings. No two marina set-ups were the same, and we would never know before we entered which berthing set-up we were coming into. We usually ran out four large fenders on either side, plus two for the stern and a roving fender if needed. Sometimes we would be berthed alongside a concrete wharf, which could be of any height, in which case we would need a different configuration of lines and fenders. In Barcelona and Cartagena, we had scored a marina like our New Zealand set-up, with floating pontoons running alongside each boat. Thankfully, these parts had minimal tides, but it could be tricky when the wind was blowing or the current was strong. Captain Bruce had mastered the bow thrusters by now, sort of, so it was becoming a piece of cake. (Ha! I can still hear Bruce laughing at this comment!)

Most marinas use the Mediterranean method of berthing. As mentioned earlier, you back your boat into your allocated berth, usually between two other boats. Stern lines are secured, and using a boat hook, another aft line is attached to the seabed forward of the yacht. This line is attached to the bow. There was no point asking the authorities beforehand what their set-up was. Our experience was that most marina managers had very limited English, and our

Spanish was sadly lacking. We had only minutes to reconfigure our fender and rope configurations before berthing: heart-pumping stuff. Entering that Gibraltar marina was quite amusing. We had our lines received by two jovial lads with the broadest English accent; it was as if we had sailed into England!

The average temperature in "Gib" was seventeen to eighteen degrees; it was so warm and sunny, no wonder the English flocked there. The Gibraltarians are multilingual in Spanish and English, but English was the primary language used. They used English pounds, but euros at the time were accepted by many traders. It was a duty-free port, so gold, fuel, perfume, cigarettes and alcohol were cheap. Glenys bought her first-ever bulk-pack of cigarettes for £32, deciding we may need to use them as currency/bartering/bribery in the Caribbean. We had never seen so many watchmakers and liquor outlets in one place – ever!

The population was around thirty-two thousand; however, the total area was tiny, a mere 2.7 square km. Gibraltar was continually reclaiming land and extending the perimeter of the "country". The main highway to Gibraltar ran right across the middle of the airport runway, run by the Royal British Airforce. At the time of writing, this was the second shortest international runway in the world. It was hard to believe Spain was just on the other side of that runway. With nearly forty thousand people per day coming across to work in Gibraltar, much work was going on. The repercussions of the Brexit deal at the time were worrying everyone there.

You could get married with only a twenty-four hours' notice in Gibraltar. John Lennon and Yoko Ono were married in the Gibraltar Botanic Gardens back in 1969 so the press could not get hold of the news.

Gibraltarians are fiercely and historically proud citizens, and rightly so. They have been famously invaded and settled by many different races and have undergone many battles over the years. The British have controlled Gibraltar for over three hundred years now and will continue to fight to keep it in their possession. Its strategic location, guarding the Gibraltar Strait's entrance, gives access to all those countries of the Mediterranean. This makes it a treasured piece of land, indeed.

We settled in at The Angry Friar pub that night, where Les and I met a few Poms who advised that the bus tour would be our best way to see the Rock. Our initial intention had been to walk it; as it turned out, that would have been a dumb idea. A belly full of fish and chips and topped up with English lager, we slept well in our new little marina haven.

Les has a magnetic force that attracts him to all, and any fishing tackle shops. Bruce's magnetic force pulls him into chandlery shops. Luckily the whole gang loved fossicking in both. The following day we took a long walk to the local chandlery. Down a very industrial road, taking us past marvellous, big-block historic sea walls and lots of military boys out for a spot of jogging…ahem! Nice. Remember, we were still on the hunt for those multinational gas fittings; no luck here either. We found a good-sized pair of bolt cutters that we needed for the boat and took off to town. Why bolt cutters? If *Marmax*'s mast snapped and fell down at sea, we had to cut it away from the hull because it could punch a hole in us and sink the boat. That was the reason, but we dared not talk about it.

We clambered into a nine-seater bus with an amusing Gibraltarian driver and guide at the wheel. Wow, wow, wow! Have any of you ever been to Saint Michael's Cave inside the Rock of Gibraltar???

We thought the Rock only had apes and intense views of three countries. What a surprise for us to enter the huge, most incredibly, beautifully preserved and presented cathedral cave system ever! It was magnificent! The cavernous cathedral had seating for four hundred people. It was used for orchestral and stage productions in the natural amphitheatre. I would not have minded sticking around for one of those nights; it was pretty breathtaking. A cross-section of one of the stalagmites was a fascination for all of us. Why had no one ever mentioned these caves to us before? I guess we never, ever imagined we'd be lucky enough to be here!

Now, onto Barbary apes. Did you know the Rock even had apes? We were amazed by how soft their little paws were! Soft as baby's hands, though the apes themselves are quite sizable. There were two hundred fifty apes within six family groups. Each ape had a code tattooed on their inner thigh identifying their family group. They were all vaccinated, so let's say they were semi-wild, hey? Although they are supposed to be wild, the government feeds them at the top of the hill twice a day so that they do not wander into the town. Everyone was warned not to feed them. The apes were very famous for stealing sunglasses, handbags and anything shiny or plastic that looked like food. One of the ladies we met in the pub was a tour guide who told us how an ape had sat on one of her client's heads and took her hearing aid out of her ear. Thinking it was a peanut, the ape ate the hearing aid. You can imagine the explanation on the insurance papers, can't you!

The views from the top of the Rock were also impressive. We could see Spain on one side, Africa on the other. Mythological history states that Hercules put one leg on the Rock of Gibraltar and one leg in Africa and pushed them apart to create the Strait of Gibraltar. Pretty cool trick!

I'd never been to England myself, but we sure got a good taste of it here. Red phone boxes, beautiful English pubs, Marks and Spencer outlets and multiple fish and chip shops on every corner with the oddest stodgy English food.

Over 70% of Gibraltarians are Roman Catholics, with Jewish and Muslim people mostly making up the balance. So much military history, magic cathedrals and robust European concrete architecture.

The crime rate was surprisingly less than 1% due to the fact you were thrown in jail for the night for even driving an unregistered vehicle. The speed limit was 40 kph, so there were no significant accidents for the hospital to deal with. The government also paid for school leavers to go to university in England if they so desired.

Glenys and I spent three long hours doing laundry for our next big trip, and we restocked *Marmax* with more stores for the grand journey ahead. That shopping trip was fun! For a start, everything was labelled in the Queen's English; we could read them. The enormous modern supermarket was loaded with all the comfort foods of home. Ever practical, Glenys found several trays of meatballs. At the same time, sweet tooth Les gleefully came running up with a full plate of chocolate Easter bunnies. We discovered two essential cooking ingredients: canned butter and an entire broiled chicken fitted neatly into a can. What a great idea, six large cans of butter got thrown onto the top of the mountainous grocery trolley, and four chickens jumped on top. With a few hundred pounds spent, we squeezed into two overloaded taxis and got delivered back to the marina to load up. All food was labelled, categorised into zones and tightly packed. This was to be our last big shop for the entire trip home.

aboard a ship, lively little restaurants and bars bursting with joyful music and laughter. We drank, knowing we would catch a taxi instead of walking twenty minutes back to the boat. As we began searching for a taxi rank, Glenys, in her usual extrovert form, struck up a conversation with an English gentleman; he happened to work for the Ministry of Defence. The dreadful mosque attack had just shaken New Zealand, and he expressed his sympathy for our home country. What he did have to say about it was interesting. Having not heard any news from home for weeks, we were eager to listen. We ended up walking alongside him for miles. It was a late night and a long walk home!

The following day, we prepared to leave Gibraltar. We had been warned to get to the fuel dock early because every Tom, Dick and Harry fuelled up to take advantage of the duty-free fuel on a Saturday morning. I guess if all the bowsers had been open, they would have had more customers! A bizarre experience was coming into a fuel wharf with an airport runway beside us, separating Gibraltar from Spain. The planes were landing almost on our deck. We finally found a live human who manned the wharf. Bruce fuelled up while Glenys decided to…escape. We needed white duct tape. Glenys also wanted wi-fi to search for a mailing address for one of her postcard recipients. The wharf was tightly secure. Glenys could not scale the fence to get into town; the tender was the only way. Janelle and Glenys set off, returning forty minutes later to a huffing and puffing Captain Bruce, no duct tape but her precious postcard at least posted!

Why white duct tape, you ask? The spray dodger enclosing the cockpit of *Marmax* leaked like a sieve. It seemingly was not designed for ocean passages but more for, let's say, Mediterranean mingling. Glenys and I jury-rigged it with 27 metres of white duct tape, so at least she looked presentable (and dry) until we could find a decent canvas maker back in New Zealand.

Off we headed, still taping up the spray dodger while in the harbour. Sails went up, and we set a course to navigate through the Strait of Gibraltar. It was simply incredible, the number of cargo ships and small pleasure craft out on such a fresh, windy day. We could not believe the dozens of small fishing boats bobbing about dangerously in the choppy seas. Thank heavens for AIS – the global automatic identification system we used to track and identify other vessels at sea.

Next came the banana drama! Now every good seaman knows that you never put bananas on a boat when going to sea. If you were not aware of this old wives' tale, I believe it as gospel. Well, *Marmax* did go to sea with a bunch of bananas; thanks, Glenys! So here we were blasting along, playing dodgems with all these massive container ships; we felt so tiny! A huge cargo ship steamed past, probably too close. He was cruising at 15 knots; we had him on our tracker. That ship had a gigantic bow wave and a stern wake you could surf on behind.

We pointed our bow, the pointy end, into the wake, as you do, to avoid unnecessary gear damage to *Marmax*. "Hold on!!" It was rather fun at the point of nosediving into the giant wave; we all yelled as lunatic Kiwis do. The gleeful "Hold on" then turned into an agonising "Oh noooooo!" Despite all of our safety drills, *someone* forgot to fully secure the foredeck sea hatch. It has a two-step system, one for ventilation, one for total lockdown security. That hatch had been set on ventilation by yours truly. How dumb. It only had a one-centimetre gap opening, but that did not stop a mountain of blue water smashing through from the forward cabin, right through into the saloon. Not only did all the winter bedding get saturated, but also the clothes, laptops, cameras, phones, shoes and anything else in the way. Arggggg!!!!! What a shambles! Did I say that nicely?

Here we were into the first two hours of a five-day sail down the Moroccan coast, and Les and I were already living in Saturation City. We rigged up a makeshift clothesline down the aft deck in the sun, but the seas had become too heavy with salt to dry anything. As a result, the interior of *Marmax* looked like a Chinese laundry on a freight train for the next week.

We entered rough sea conditions with huge swells and 20–30 knots of wind blowing. Fortunately, everything was dry by the time we got to the Canary Islands. The first meal cooked on the trip at sea, I whistled up a heap of spaghetti bolognese. Lots of delicious tomato sauce, you know what I mean. Serving such deliciousness in a lurching sea is no mean feat. One must steady oneself by jamming your body in a place in the galley, sometimes halfway up the wall. You must then attempt to juggle pots and plates on a gimballing cooktop in symphony with the violent lurching of the boat. An unpredictable wave slams into the hull. As you can imagine, I managed to spill hot red sauce all through the tracks of the cupboards and over the walls while serving up. The omelette mix was dumped in the same place the following day while we copped another big hit to the hull. Oh mannnnn…We sat Bruce and Les down and watched those bananas get eaten with gusto. Our dumb fault. Damn those evil bananas!

Many people at home would have been getting a good laugh while following our tracking. We had to tack (zigzag) down from Gibraltar, past Tangier and followed the Moroccan coast. We had been advised to stay 20–30 miles off the coast to avoid working fishing boats and their nets. We had reasonably consistent 20–30 knots abeam by now, with massive rolling swells up our stern all the way. The wind was roaring out that funnel, out of the Mediterranean and down to the Canaries.

For the more significant part of this trip, we were on autohelm. I can't emphasise enough the fantastic capabilities of this remarkable equipment. If the sails are set right, it can take high winds, surging and rolling seas and the violent twisting that generally goes with it. It adjusts your steering automatically to keep you on course. When things got hairy, obviously, we flipped back onto hand steering. To slow the boat and make sailing more manageable, we would reef down, i.e., we made our mainsail smaller by winding some of the sail down along the boom. Similarly, we could roll our headsail in to make it smaller too.

Our speed was not compromised when we did this; it made things easier on the gear and more comfortable. We maintained 6–8 knots for most of the time, often doing 8–9 knots. Les held the record with an 11.3-knot run. We needed a scoreboard; it was getting competitive now. We were never to hear the end of it from him. The days were getting longer, and the further south we went, it was getting warmer. We had settled into our routines of two-hour watches, which had worked well. We slept/read/wrote/did hobbies/Janelle's schooling and stuff…amazing how busy we were!

Janelle was in year 10 at college in New Zealand. Here she was, floating about out at sea, hoping she would get away with zero lessons for a year. Les and I had agreed to home-school her early on in the piece. I tackled English, Les taught mathematics. Let's be clear, we are not schoolteachers, but I don't mind saying, we sure had to learn in a hurry! With a strict teaching curriculum, we churned through those lessons through thick and thin. Janelle started off a bit slow, but she persevered and became quite a scholar at sea, to her credit. Something essential on long passages was to have a structure in our days, coupled with goals to attain along the way. Try driving your car along a desert road for five, ten or twenty-three days straight. You'll get my drift.

We had year 10 workbooks Glenys had to source herself, as the New Zealand school system did not accommodate offshore students. You see, we had no wi-fi for schooling; however, we did for boat emergencies. We could ill-afford days on anchor in foreign countries just to hook up with the technology required by the method of correspondence teaching. A new country meant real-life geography and biology lessons. This was the most perfect way to complete a year of school when you were fourteen years of age! Exploring spooky caves, peeking into lava holes and volcanoes, patting apes, snorkelling with sea lions and dolphins, cartwheeling across exotic beaches, climbing lush mountains, and throwing your head up under fairy-tale waterfalls.

The flipside was lessons on a roller coaster yacht can be fun. We would jam our knees into the galley table to brace ourselves for the next slam from a rolling wave. Pens and pencils would go flying. The cupboard full of schoolbooks would randomly try and knock us out. Water would rush past the windows behind Janelle's head. Water would be swirling and foaming over the decks above our heads. She had to learn to concentrate. There was always a power struggle going on in Janelle's head. Reading books seemed not quite as attractive as watching a movie on a laptop, for example. Each day, at the start of lessons, we made her yell out, "I'm alive, I'm awake! And I feel great!"

Try it when you wake up in the morning: "I'm alive, I'm awake! And I feel great!" No, louder! Throw your arms up in the air. You'll be surprised by how energised you will feel. Next time you are faced with a yawning teenager with wild hair at the breakfast table, I can recommend this invigorating chant!

After five days and four nights, we made it safely to the Canaries. Not a biggie trip, but long enough for Glenys and Janelle. They previously had a history of seasickness. Actually, we did not know how they were going to handle this new boat. God knows how they would take the next stint on the Atlantic crossing. It was blowing quite a bit as we entered Arrecife Harbour at Marina Lanzarote. *Marmax*'s decks were washed down with fresh water, and the interiors were given a spruce up. After a quick fix of wi-fi to family, we then had a race to the brand-new marina showers. Oh, to wash the sticky, stiff salt out of your hair after seven days at sea, sheer bliss! It was so far, so good for now.

The Canaries were nothing like we expected. Over their long and eventful history, the Canaries have been called the Elysian Fields, the Fortunate Isles, the Garden of Hesperides, the Enchanted Islands, and the Eternal Spring Islands. They are an archipelago of eight high volcanic islands that rest on the Atlantic seismic ridge, reasonably close to Africa. It is a strange landscape. They have sandy deserts, stark lava fields, lush mountain valleys and a snow-capped mountain peak, a dormant volcano rising over 3.7 kilometres off the ocean floor. The water depth drop-off around the islands was quite incredible.

As we entered the harbour at high tide, we noticed four people swimming just below an old stone fort on the rocky shoreline from a distance. In the back of our minds, we thought it was a bit chilly to be swimming. Turns out, this was a fascinating piece of bronze art. Low tide revealed four bronze statues of men astride horses. Very smart and hugely impressive!

The islands are steeped in ancient history. Christopher Columbus set off from these islands on three of his four transatlantic voyages, and many of the buildings of that era still stood. Over a thousand yachts use the Canaries as a springboard to get across the Atlantic to the Caribbean every year, an example set by Columbus over five hundred years ago. And duty-free: a bottle of beer was only €1! I won't bore you with all the detail, but it sure is an intriguing place, still Spanish, with the euro currency.

Lanzarote is around 846 square km and only 125 kilometres off the north coast of Africa. We were berthed in the port city of Arrecife. The town, full of castles and forts, authentic Spanish bars and delicious seafood restaurants, encircled a tidal saltwater lagoon dotted with dozens of fishing boats bobbing on small moorings. You could stroll right around it; the water was sparkling. One afternoon, laden with more shopping, we were trudging home to *Marmax* like baboons with our knuckles almost scraping the ground. The loads were heavy. We decided we would take a shortcut through this square courtyard paved in small volcanic pebbles. Midway across, we looked down at our feet, wondering why the short journey not only stank like nothing on earth, but we appeared to have doggy-do stuck to our shoes. We were out of there! We had not seen the prominent signs on each quarter of the courtyard; well…they were in Spanish! It was a reserve for dogs to do their business and a "cigarette park"; somewhere to mash out ciggies! We were bent over, laughing so hard, with tears streaming down our faces. Man, we were dumb. Why have such a disgusting thing in a quaint little tourist town?

Meanwhile, back at the boat, Bruce was in one of his dark moods. It seemed that the SD card, which held the Navionics chart, a marine chart we needed to find our way home, did not fit our plotter. Glenys and Bruce walked for miles and miles, trying to find a computer technician to help. They eventually found one, but he could not copy the file; remember our problem of not speaking Spanish? Bruce would need to buy new electronic charts or copy some onto a smaller card. Wait for it, rookie error 3: a lightbulb moment from Bruce lying on his bed, at his wit's end, staring at the ceiling. The little SD card we needed was housed inside the big SD card carrier he was trying to put into the plotter! Problem solved. We exploded with high fives all around and sighs of relief. The second cause for celebration that day was the gas fittings were fixed! We were then able to handle multinational gas bottles. Bruce finally had a smile on his face...

The port of Lanzarote relied heavily on visiting cruise ships and had spent millions gearing up for these big floating beasts full of cashed-up tourists. Our marina was set up as part of the cruise centre. As the bland-faced tourists sauntered off their ship, they were herded past the marinas to check out the colourful array of overseas yachts. This was a rather impressive sight, I must say. We flew our two New Zealand flags with pride. Our big silver fern flag flying from our mast made it easy for us to find our way back to *Marmax* wherever we went. It started conversations too and stuck out like the proverbial.

The cruise ship passengers were then hustled through to the local street markets, geared up especially for the crowds. There was one big thing that had struck us about all of these ports, which catered for the cruise liners. Why were all the shops and markets the same in every town and every country? Outside of our marina arm, we had Gucci, Versace, Kenso, Pandora, all duty-free, but not one customer in them.

It was so much better to travel on a private yacht. We felt truly blessed every day with the freedom to chase the sun, do whatever we wanted, and when we wanted. Sounds blissful, doesn't it? Anyone who has sailed the seas knows it is not quite that easy...

After two nights on the marina, we sailed from Lanzarote down to Gran Canaria. The ocean was up to four kilometres deep in the channel between islands. While we had been in the Canaries, it had been blowing a constant 18–20 knots, making for some excellent sailing aboard *Marmax*. On the way to Gran Canaria, we passed approximately 30 kite surfers off one of the distant beaches. The landscape looked bleak from the sea: grey volcanoes and acres of desolate lava rock. Then we came upon those colourful kite surfers skimming magically across the water, in the middle of, seriously, nowhere! Weird.

Much to our frustration, it took us twenty-four hours to contact the marinas in Gran Canaria to raise someone to help us get a berth. They simply would not answer their VHF radio. We managed to obtain permission to stay just one night moored to the fuel berth. We were told that another fourteen overseas yachts were waiting for a berth there. Our luck continued. Maybe we were getting favours because we had a cute kid on board!

Unfortunately, the marina was so full, there was no room for us to stay for a second night. We had hoped to do a bit more exploring. This was the off-season for yachting here; Lord knows how everyone got on in peak season!

Unbelievably, over two million people live on the seven islands of the Canaries; most people live in Gran Canaria. Before arriving, the mere name of the Canaries had suggested to us delusional Kiwis, islands of lush growth and tangerine trees teeming with bright yellow tweeting canaries. Though this may seem logical, they actually get their name from the Latin term Insula Canaria, which means Island of the Dogs. The ancient Romans, who first visited the islands, gave them this name – so disappointing!

Although considered a romantic holiday and cruising destination, we were not sure about Gran Canaria's rave. I believe it had lost its way a little over the years. Still Spanish but lacking the authenticity of our previous travels. A bit of a melting pot of different cultures may have been the reason. A quick excursion ashore revealed a kaleidoscope of varying architecture with no clear sense of the authentic culture. It was nowhere near as beautiful as other islands visited, and the streets and pathways were quite littered. One could not help but notice the security fencing and cameras everywhere. If we had more time, we were sure we would have found some cultural substance in the hills, but such is life.

We started our final preparations on deck for the long haul across the Atlantic. The boys rigged up the emergency sea anchor and double-checked all of our safety gear. We were heading off in the morning. As an English lesson task, Janelle wrote a blog describing what safety gear we had on board and how it would save our lives in an emergency. We imagined family and friends would have felt quite relieved to learn about the multitudes of safety equipment aboard *Marmax*.

We were about to be kicked out of port, yet we still had a few critical tasks to complete. We threw on our backpacks and walked three kilometres to the closest supermarket. Bruce, unable to find a taxi, departed on a five-kilometre hike to find customs for our clearance out of Spanish territory. What drama! I won't go into too much detail. Just as we were at the supermarket checkout, laden with, yes, more groceries, Bruce called to say that we needed to join him at customs immediately. This would involve hauling some weighty loads of groceries for miles in the heat, as we could not find a taxi, and very few people appeared to speak English, which was no help. Apparently, the customs officials needed to eyeball us before clearance, which they did. Laden with those grocery bags, we eventually made it back to the marina several hours later than intended.

We rushed for our last luxurious hot showers on land, the groceries were stowed safely, and blogs sent; lines were finally thrown at 1.30 pm. Sadly, we couldn't stock up on alcohol due to our lack of land transport and our inability to carry 100 kg each through the streets. "We can do without it; we have enough for twenty-five days." Famous last words they say! Sure.

We were now looking forward to finding a little island paradise in the Caribbean sun, full of rum, and expecting to get to St Lucia around April 16, some 2700 nautical miles away!

Sailing Canary Islands

Trying to enter the port of Gran Canaria!

Gran Canaria

Lanza

6. Who Said the Atlantic Crossing Would Be Boring?

As the last of the land disappeared behind us, good old Les says, "Gee, we haven't seen rain since we left Spain. Wouldn't it be nice?" Weirdly, within thirty minutes, it bucketed down entirely out of the blue! A quiet, gentle night of motor-sailing due to lack of wind followed, as the last of life as we knew it, on land, disappeared behind us.

Heading south, we continued down the Western Sahara Coast to, hopefully, catch the trade winds and scoot east, just above the Cape Verde Islands. Destination St Lucia, Caribbean Sea.

We were becalmed a few times on the way but just had to sail our way out of it, as we held only seven days' worth of fuel aboard *Marmax*. The rigging clanging on the mast, sails, and gear truly drive you nuts when there is no wind; I can only imagine what it would be like being stuck in the Doldrums. Every little sound was magnified. The clunk of the boom vang bouncing in its gooseneck attached to the mast, the whacking of the sails desperately trying to fill, like dying fish on the deck.

Our incredible thrill during these calms was, hearing from afar, the gentle breath of dolphins as they bounded towards us from all directions. "Dolphins!" was the cry, and all crew leapt to action, clambering sleepily from bunks, seats, homework, reading, whatever, up the galley steps to chat away to these remarkable mammals. Compared to their Mediterranean cousins, the Atlantic dolphins were much smaller in size. Their bellies were a light grey with speckles all over them. They also did not appear to have scars on them, leading us to speculate on a much more isolated existence.

One hundred and twenty miles off the Canary Islands' coast, we almost sailed straight into a bundle of floating white marker buoys with a flag on top of each one. There were no radar reflectors to be seen, and it was suspended in 4500 metres of water. It seemed an impossibility to think it may be a fishing net. We decided to shoot through, away from it. We had cooked up a juicy little story about it possibly/probably being a drug pickup set up; what else would it be? There were still a few container ships about. Very interesting on an otherwise uneventful day.

The following day, the wind had picked up to 8–10 knots, still slow going, but at least it had an energy about it. Imagine our excitement when a pod of some twenty orca whales cruised past us in the opposite direction, their stunning black and white bodies translucent in the sapphire blue water. A lone turtle drifted by, along with sightings of dozens of shining man o' war jellyfish sails and another five groups of dolphins during the day – indeed a time to feel alive! The Atlantic is so utterly unreal in its blueness; we often took our sunglasses off just to enjoy the pure intensity of the colour.

One ink-black night, we were sailing along, making good time in light winds, when *Marmax*, for no apparent reason, slowed up and almost stopped, even though the wind was still blowing 6–8 knots. It had us stumped. Had we become entangled in a fishing net or a mass of weed? Hastily, we flicked our spreader lights and torches on and examined the hull. *Marmax* remained stubbornly stationary, with the sails and rigging getting angry in the heaving ocean rolls. Slam! Slam! Nothing could be seen at all. Since we could not move forward, we turned on the engine and threw her into reverse; no feeling of release, still nothing to be seen. We throttled forward. After a couple of minutes of backing, we were free. No rhyme or reason, but we were off on our merry way, thank God! It is not a good idea to overthink these things!

When you are at sea, one of your main entertaining activities is food! We looked to our fearless hunter, Les, and hinted that another tuna would be kind of nice to catch so we could make sushi and sashimi. The lure had been trailing for most of the day and into the early evening. Then the piercing scream of the reel, from the fish on the line, went off like a steam train. Oh geez, I thought, I hope it's not a marlin; I saw a fish leap high out of the water about two hundred metres behind us. When sailing a yacht under full sail, even on a reasonably calm day, to drag a fish in of this size on a rod and reel set-up with braided line is quite a feat. You can't stop the boat, but you can slow it down a little by pointing into the wind, releasing the pressure in the sails. A fifteen-minute fight, man against beast, and jubilant ye-haaing as Les and crew landed a rather spectacular northern bluefin tuna, otherwise known as chicken of the sea. Les filleted the fish on a heaving, slippery deck. At the same time, we girls acted like nurses: dealing with all the excess blood, buckets of water, the handing of the gaff and knives. What a team! That tuna was divine. Soft as butter, delicious for seven meals feeding five people.

The wind had been steadily picking up for the past five days. It had become almost impossible to crawl across to reach the back deck, let alone throw a line over. Les would have to wait another day for his next fix of adrenaline. In the meantime, Janelle had been crowned master chef of sushi. Delicious!

Bruce surprised us; on our tenth day of the passage, he burst into song! He had dug out 'his' music on his tablet. He proudly connected it to the Bluetooth speaker in the cockpit. The title: Beer Drinking Songs. Headbanging favourites from his heyday. Hmmm…who would have imagined?! We often sang all day long, sometimes in unison, more often not. Bruce was probably sick to death of listening to us and our repertoire of Spotify *Marmax* songs. "Knee Deep" by the Zac Brown Band had become our official onboard anthem. Glenys had carefully transcribed the words so that Bruce could join in.

The trade winds were now with us, with steady breezes and an established wind pattern of 15–22 knots. The days had begun to blend into each other. The sailing, in the past few days, had been phenomenal. You could smell the sea and feel the air as *Marmax* powered through the sparkling blue water. Some pretty heavy rolling swells, with breakers crackling beside us, kept us on our toes – literally. By now, we were all collecting plenty of mean bruises. We got thrown into bulkheads, jammed in the fridge and bathroom doors. Our hips were getting hurled unexpectedly into the galley table and stove. Les's head connected with a fire extinguisher one night, and I went airborne in bed, landing with a great whoosh from Les's lungs as I crashed on top of him. The boat interiors were getting salty, making the floors dangerously slippery.

The art of carrying two cups of hot coffee up to the cockpit was probably the most celebrated talent you could have on board at that time. A double rum and Coke was on par, to say nothing of whipping up meals on a bucking horse with flames dangerously close to your belly button all the time!

As advised, we took turns doing two-hour watches. I was lucky enough to do the 12 midnight to 2 am shift, easily one of my favourite times of the twenty-four-hour routine on board. The skies were indeed a surreal and ethereal sight, with stars and planets laying on a superb display of magnificence, night after night. As we rocketed through the water on these clear, luminescent nights, you could feel the rhythm of the earth spinning as the sun settled. The moon rose, and the most incredible things would happen. My personal biggie came on such a night, one of the most amazing experiences in my whole life.

It was a relatively calm sea. A lovely 10-knot breeze was pushing *Marmax* quietly through the water under full sail. The only sound was the waves bubbling along the waterline, glowing in the phosphorescent light. Suddenly, I heard the distinct sound of a school of dolphins galloping through the water from behind us from far away. In turning, seriously, my eyes nearly popped out of my head. I was not sure how many there were. Twenty to thirty dolphins surged in a big frothing phosphorescent triangle of bubbles. Zooming in behind our stern, they split into halves up both sides of the boat, giving the most magical performance I could ever hope to witness. Gasping softly, their smooth bodies rhythmically surfaced. Glowing bubbles and iridescent mercury balls were trailing deep under the water, cutting ribbons through glassy water. It was as if *Marmax* was sailing through the Milky Way...

I could not yell out "Dolphins!" to the crew, as we would have during daylight hours, as they were all sleeping; well, trying to! I could not go down and shake anyone awake for fear the dolphins would leave while I was below. They usually didn't stick around for too long. I just could not believe it. Tail thumping and pirouettes across the tops of the waves, some of them looking up at me with one eye as they saw my face in the dim navigation lights.

They flew alongside the hull for around five minutes, leaving in a formation to port, joining together as they gleefully took off. I just slumped back behind the wheel, leaned right back to look up at the stars, and shook my head. Did that just happen? Who said these long ocean crossings were boring?!

By the sixteenth day of our Atlantic crossing, we were all getting a little tired of the constant rock 'n roll motion and having our meals served in a bowl. We could not do plates; everything slid straight off them! And the broken sleep…But we were eating very well, and everyone was in good spirits. There was a great vibe on board, and Janelle's schooling routine kept us all amused and focussed.

I could again paint a picture of sheer bliss here; however, life is not always perfect on such a voyage. We did have things go wrong, which required calm heads to fix. Emergencies, usually in the dead of night, where it was "all hands on deck." On another of my witchy watches, we experienced a bizarre occurrence where we appeared to have sailed into some sort of wind vortex. It was after midnight, blowing 20–25 knots, with a following sea and not a calm night at all!

Screaming wind in the rigging, big rollers from behind, and random punches left and right on the hull from the uneven swells. Suddenly, a complete wind change hit us from the opposite direction. We had two reefs in the mainsail and the genoa (front triangle sail) furled right up to a storm-sized sail. *Slam!* went the boom as it raced on the traveller across to the other side of the aft deck. Everyone leapt from their bed and scrambled for life jackets and lifelines to click on before entering the cockpit. Geez!

"Gybe!" Bruce screamed (to go in the opposite direction). We whipped over to port (left), ropes whacking, mast trembling, winches screaming. Again, "Gybe!" Opposite direction. Hold it! Hold it! "Okay…Gybe!" Oh boy, we slammed around four times until she settled. What the heck was that about? We would never know. It was merely the fickle winds of the mighty Atlantic! Thank goodness *Marmax* was a strong lady with a sound hull.

The brand-new furling rope on the genoa had almost chafed through due to non-stop sailing under full load. That had to be replaced. The first reefing rope on the mainsail had rubbed through and broken. Not sure how we'd fix this one, as it was too dangerous to get near the boom in this bucking sea. There were always running repairs going on, as the wear and tear of the ocean's constant movement took its toll on the gear – let's say there was rarely a dull moment!

Glenys announced we would be out of milk in a couple of days. Great! We should have thought about that before making custard squares, cheesecake, muffins, and scones, scrummy as they were! A more evident crisis: we were almost out of alcohol!! We were also down to only a few potatoes, onions, and a lonely cabbage. Our creativity in the cooking department became exemplary. There was no meat, lots of pasta and rice, and a few non-descript cans of "stuff" on top. Most of our canned and packaged stores were in Spanish; thank goodness for the pictures on the labels. Others had questionable contents, so it became a bit of a guessing game. There were several legitimate reasons as to why we had run out of stores aboard. But too many reasons to list here! Let's say it's politically correct to stay mute!

Suffer, little children. Les needed to catch us another fish, but the wind and the swells were still too high to enable him to safely do this. *Marmax* was moving through the water far too quickly to put the line out. Besides that, there were increasing numbers of large seaweed mats floating by. This was Sargassum seaweed. Starting life in the Gulf of Mexico, the stringy, brown seaweed is pushed by sea currents out to the North Atlantic, then across to the Caribbean.

Les was so frustrated about the stuff; he had to make do with flying fish crashing at his feet in the cockpit at night. Flying fish would zoom off the crest of high waves and land with a soft thump onto the deck or right into the cockpit. They are rather beautiful, and we saw more and more of these wondrous creatures as we approached the warmer waters of the Caribbean.

"Can we eat them?"

"Nar…too bony."

A bird landed in the cockpit one night too. Lucky it escaped without getting its legs crumbed.

One unusually warm day, we had a very gentle breeze blowing around mid-morning. Our laundry was strung up over the deck; it was just gorgeous. Glenys and Janelle sat on the stern of the boat, trailing their feet in the water to cool themselves down. They begged Bruce to let them go for a swim; we all wanted to. Just as we were ready to pull down the sails and hove to (stop), we saw a large ship on the horizon come into view. If we were to stop our tracking across the water and pull down the sails, it would probably look like a distress signal to the oncoming ship. We were also confident they had us locked on their radar and could see us. The last thing we wanted was a 900-foot cargo ship pulling up alongside to check us out. So we waited patiently for the vessel to pass; this took about twenty to thirty minutes. The trouble was, by then, the moment had passed.

The wind had increased, and with a sigh of sadness, we resigned ourselves to the fact there would be no swimming in the Atlantic that day. Bummer.

The enormous floating carpets of spiky, yellow seaweed mats were becoming more frequent now. Sometimes they were up to ten metres wide and fifty-plus metres long, streaming along the currents. The seas were still azure-blue, stunning, sparkling clear waters, just so glorious. The stars at night were so dazzlingly bright, the sea appeared to always to be bathed in moonlight. We all did stargazing and compared notes the next day.

Les's most precious night-watch story involved shooting stars, of which there were many. As he was gazing up to the skies, a burst of light flashed in front of his eyes. Zap! Gone! "What the hell was that?" He scrambled to the radar and zoomed in with the scroller to see if he had overlooked the sighting of an approaching ship. Perhaps it was a flare he saw? A meteor? As he stared again, he was mesmerized by a series of five shooting stars over twenty minutes, all from the same quadrant. This got him thinking. That dynamic and bright light he saw only momentarily must have been a shooting star, shooting out from the spot directly at him. Les has always wanted to be a space traveller; he believes he was born two hundred years too soon. This was a thrilling life experience for him. He still talks about it!

We were three days out from the Caribbean by now, after nineteen days at sea. There was a bit of excitement on the deck as we saw the first aeroplane jet streams overhead. A family of mothers and their baby dolphins thrilled Janelle up on the bow, some of them only eighteen inches long! Each day blended into the next; routines had become massively important to our sanity aboard.

Day 23, we were now 70 nautical miles away from our next destination. Because of prevailing wind conditions and the best point of sail for the boat, we had decided to change course from our original destination, St Lucia, which is in the south of the Caribbean. Our sail plan was changed to head towards the island of Virgin Gorda, one of the northernmost British Virgin Islands in the Caribbean.

Glorious, glorious sailing. We had experienced a steady 20 knots for the last several days. We broke the second reefing line on the mainsail and had to continue as best we could. We now had two reefing lines out of action.

Glenys and I had a ripper of a rainstorm early one morning, which pushed us along with an exhilarating ride. Suddenly it dropped right off, and we reverted back to the painful slapping of the rigging. Never to lose an opportunity, Glenys coerced Bruce into dropping the sails so Janelle, Glenys, and I could go swimming. Oh wow! The warm water was like jumping into an effervescent bubble bath! From in the water, it was a bit alarming watching *Marmax*'s big hull smashing in the ocean swells, rising and falling some three metres at a time. That would have been a big donk on your head with five kilometres to the bottom. We had floating safety ropes out; it was pretty exciting stuff for us three girls, anyway!

By now, we were feeling a little sad to be leaving the mighty Atlantic. We knew we were about to have our final night out there before getting into Virgin Gorda. To have had twenty-three days of total isolation and peace from our usual busy worlds was indeed a privilege. So much time to think clearly, dream and truly relax, no emails, no phones; you can imagine it can't you…sheer bliss! The last night at sea, totally without coffee, alcohol, or ice, was the only heart wrenching reality in our tiny world!

The ocean depths went from a heart-stopping six kilometres deep to a very shallow seventeen metres as we entered the Caribbean Sea. The island of Anegada appeared on our port side. We weaved our way through dozens of tiny white fishing buoys while hard on the wind. The landscape looked amazing as we approached. The sea was as Caribbean blue as you can imagine. After 3300 nautical miles, approximately 5280 kilometres, we arrived in Virgin Gorda, the British Virgin Islands. Time to party!

7. Caribbean Time!

"You are not dreaming…You are in paradise!" I found this written on the side of a boat here in the Virgin Islands, and how true it was!

Let me tell you about our first experience entering a Caribbean port after spending those twenty-four days at sea. To begin with, port and starboard navigation buoys are all back to front there. We had previously learnt about it, so there was no problem. We decided the first landfall would be Virgin Gorda Yacht Harbour in Spanish Town on the island of Virgin Gorda in the northern region of the Caribbean. We had to clear customs and immigration through a clearing port; this seemed as good a place as anywhere.

After being out at sea with never-ending horizons of blue for so long, it was kind of nerve-racking to navigate into this tiny harbour. The wind was blowing a hard 20 knots when we entered the marina. The water was so clear; it was as if we were floating on a sheet of cellophane. Below the boat, everything could be seen: vicious-looking coral, dark jagged rocks, and colourful fish swimming about.

We held our breath…our keel must have skimmed only a couple of centimetres above that coral beneath us. We made a tight turn to port, just a few metres from waves frothing over a small reef, then popped out safely into a compact one-hundred-berth, full-service marina development. What a surprise! Another fast turn to starboard and a sharp and shiny, uniformed marina attendant appeared, looking rather dashing, standing on the end of the marina arm to catch our lines. These marinas were not floating affairs; they were built for hurricanes and made of solid concrete and steel. We were instructed to reverse in on the end of the first arm. Did I mention the wind strength?

Up we nosed into the wind, sideways we got swept. After weeks at sea, it really is terrifying to carry out a manoeuvre like this; that concrete was foreboding, our hull was precious, and the wind was strong…Thank God for bow thrusters! As I mentioned earlier, all this stuff happens very quickly. As usual, I was on stern lines. Glenys was on the central spring lines, Les on the bow. The guy was standing there, looking very casual, VHF in hand.

I yelled loudly, above the wind to be heard, "Could you please grab our spring line first?!"

"No panic, lady," he crooned in his smooth, husky voice. "We're on Caribbean time!"

In our minds, we were screaming, "Catch the damn rope, man! Our yacht will be smashed into your fancy concrete marina!" Having researched island etiquette before arriving, we were politely silent.

"Spring line, ma'am," he breathed with one large hand held out.

Glenys threw the rope. The line was swiftly wound around the marina cleat. The wind had us, yet miraculously we slipped very quickly alongside with a very gentle thud and came to a stop. Seriously, we had only five centimetres to spare on the stern; I heard

Bruce's lungs whoosh a great sigh of relief. A perfect landing without the dreaded crunch.

In the next few days, we discovered that the handling of boats here was quite an art, really cool to watch, and very classy. Even women and children are seemingly unflappable. All manoeuvres were carried out at high speed, mainly because strong trade winds seem to be prevalent here; I guess they'd had plenty of practice. Deckhands would stand up straight above the marina cleats with the lines, flick, flick, flick…It was all in the wrist work. It was pretty funny to think of how freaked out some people get when going into marinas back home! Welcome to the world of charter boats, or should we say the Kingdom of Catamarans! Easily 80% of the vessels were catamarans here, with the balance being yachts. All shapes and sizes, not many small ones, and probably mainly between 40- and 80-foot lengths. There certainly was some magnificent, mind-boggling eye candy floating around here! Money was no object for most; everyone was there for a good time. It was another world, I kid you not!

As we got our salty swagger happening, the first thing that struck us was the sheer devastation ashore; the island looked like a war zone! For those who are not aware, the Caribbean was hit by two category 5 hurricanes within two weeks of each other in September 2017, Hurricane Irma then Hurricane Maria. The properties were annihilated, the people devastated, poor things. These two hurricanes combined killed approximately 3200 people. Twisted steel, trees ripped out, empty facades on their homes and businesses, dozens of wrecked yachts and catamarans scattered in the boatyards. Some companies managed to carry on with the help of painted marine ply and creative landscaping; a few pop-up shops and restaurants had set up under makeshift tarps. You did have to admire their resilience.

The boatyard manager told me that the winds had gotten up to 302 kilometres per hour during the hurricanes; the catamarans had started spinning when the high winds lifted them. They were flying around in the sky, some landing four deep on top of each other. Remember, these were huge 40- to 80-foot-long luxury catamarans there; it must have been dreadful. Rumour had it that approximately sixty-five thousand vessels from Florida to the Caribbean were destroyed during this time. In the coming days, we were to see hundreds of boat wrecks scattered and smashed upon the rocks and hills as we travelled around. It must have totally ripped their hearts out; the place was wasted.

The second major standout was the beautiful, friendly Caribbean people with teeth like piano keys beaming out at us. Smiling, kind, warm, and softly spoken. Why couldn't the whole world be like this? I guess it had something to do with being on Caribbean time.

People often asked us about our sea legs after sailing for weeks on end. How was it adjusting to walking on terra firma? A funny thing was we would jump onto the land, straight off the boat, and go for a walk, no problem. No falling over, no dizziness. However, it was hilarious going into the first tiny local shop in Virgin Gorda. The first stop was, of course, the chandlery. It was a makeshift building, patched up after the hurricane damage it had copped in 2017. The low white ceilings and narrow aisles made for excellent entertainment with our body balance for some weird reason. "What shall we do with the drunken sailor, what shall we do with the drunken sailor…"

If you caught a taxi in the Caribbean, you could be waiting in the sun for another twenty minutes before heading off. The driver would wait until another paying customer turned up. For the uninitiated, this custom was so very frustrating! They charged by the head. The taxis were various-sized little trucks or utes, seating between six and twenty people, obviously depending on the size. We had one particular taxi man recommended to us, Andy. He had a lovely bunch of colourful bougainvillaea tucked in the back tailgate and soft vinyl seats; that was good enough for us. It turned out that Andy also owned one of the local beach restaurants, Fischers Cove. We came to a conclusion, reasonably quickly, that this taxi business was his cash business.

When you ask a Caribbean taxi driver, "Where is the laundry or supermarket, please?" the answer will always be

"Awww…fifteen to twenty minutes that way," with a toss of a banana-sized thumb. Invariably, that laundry or supermarket will be less than four minutes away, and they just managed to score a US$20 note from you to get there. This happened a few times to us. We got to the first laundry to find nine washing machines and five dryers inside. We assumed it used to be a small shopping centre, but it now resembled a bombed-out ablution block.

"Sorry, lady, only one machine working," said this gentle-mannered man with a toothy smile. He cocked his head and stared at all five of us lugging huge heavy bags of laundry then shifted his gaze to the other men waiting for that one machine. We were out of luck here; sweet smile though, thanks, mate. Dammit. We lugged the laundry back out to the big grinning Andy.

"You go somewhere else?" He said with his hand on his heart, "I take you!"

Stuff it, we were going to have to lug the laundry to the supermarket with us and grab some stores for *Marmax*. Into his taxi, we all clambered. Another five minutes around rickety narrow roads with chickens scattering, past a dump full of broken houses, boats, and cars. We handed over another US$20 note.

"I wait for you; you leave your washing in my car." Grin, grin.

We were up to these tricks. Would this cost us another US$20? More than likely. Would we ever see 80% of our clothing and bedding ever again? Questionable. To keep it safe, we offloaded the laundry onto the road and left Bruce to stand guard. After a quick wander around the battered and bruised two-horse town, we blew off a few hundred dollars at the supermarket, hauled our groceries out to the street, and piled them alongside Bruce and the laundry bags sitting in the sun. We called our mate Andy back; better the devil you know than the devil you don't. Andy was having a great day! Our ever-growing load prompted us to head back to the marina. We'd had enough of this. We were not used to walking, for a start, and certainly not used to the beating heat radiating out of the stone roads.

We had met another fellow yachtie earlier on in the chandlery shop, Canadian Vanessa. She saw us arriving at the marina with Andy and got stuck into him after hearing of our laundry dilemma. She told him to take us to this particular laundry; she seemed to think he should have taken us to it in the first place. You know what? Andy was smart. He probably knew this all the time. I bet he was having a good giggle inside. Glenys and Bruce jumped back into Andy's money-making machine, and off they went. Les, Janelle, and I trudged back to the boat, exhausted, and loaded the groceries into *Marmax*.

A few hours later, a very weary Glenys and Bruce returned with half the laundry washed but not dried. The other half stayed the night at the laundry. The laundry lady had put powder detergent at the bottom of each load, so the laundry powder lay caked to the bottom of the machine; it had not dissolved. All the clothing, sheets, etc. smelt like cigarettes, not lovely, fresh fragrances! They returned to pick up the second half the next day and the same deal. Everything reeked! By the way, none of us smoke.

We ended up hanging up all of our washing on *Marmax* to port (left-hand side of the boat) so that we did not look like a bunch of loser hippies in amongst all this class. All the girls' "little bits" were carefully hidden on the lower lifelines to not draw attention. Just our luck! A big tourist boat pulled in alongside our port side with around forty Americans on board. Kind of embarrassing, though a few of them were a curious sight too. Brown voluptuous girls in skin-coloured bikinis strolling down the marina will always make your jaw drop!

During our three fun days in Virgin Gorda, we managed to clear customs. We spent hours cross-legged in the dirt outside the ice shed, the only place we could find wi-fi, catching up on communications. We then managed to put a big dent in Bruce's credit card as we purchased an American adapter for our English and Spanish power outlets for shore power. We partook in the most decadent piña coladas and bush whacker cocktails you could ever imagine. The Bath and Turtle Restaurant, Fischer Cove, and the stunning CoCo Maya Restaurant and Bar on the beach in Spanish Town were all superbly Caribbean. All beaten up, boarded up, no windows, but full of fun, laughter, colour, music, and tropical deliciousness!

One afternoon, an official-looking lady in uniform was on the marina at our stern taking notes on a clipboard. I asked if I could help her.

"Hello, I am your security guard for the night." Big smile. That's interesting. Why did we need security? Ahhh…Easter weekend! At 3 pm, the Puerto Ricans started arriving in their magnificent luxury launches and catamarans. The party people had arrived!!!

We worked this out: If you were Puerto Rican, you must have

four 350-horsepower outboards on the steran of your muscle boat for every eight children you have. We also adored the names on them! One speedboat owner had six daughters; their names were all written on the stern. Puerto Ricans appeared to breed many daughters, and, oh boy, did they LOVE their music!

Middle-aged men strutted along the waterfront esplanade. Bare-chested and tanned, they wore thick gold chains around their necks. Gorgeous women paraded along the marina in itsy-bitsy bikinis, see-through crochet beach tops, and teetering designer heels. Kaleidoscopes of coloured lights pulsated to the rhythm of the beat from under the waterlines of their boats. What a fabulous atmosphere! The music went on until the wee hours, and a gentle tropical breeze blew through our open hatches. We enjoyed a cocktail-induced sleep with smiles on our faces. Yes, siree, we were going to love the Caribbean!

A little background information for you all…the abridged version! The British Virgin Islands are an archipelago of around sixty stunning islands, coral cays, islets, and rocks. With endless white sandy beaches, they are usually fringed with green forested mountains, though when we were there, they looked like they had been hit with a flame thrower! Each island was amazingly unique. The BVIs sit around the centre of the chain of islands that make up the Caribbean.

Christopher Columbus claimed the islands for the Spanish in 1493 and named them the Virgins. The Spanish found copper on Virgin Gorda in the 1500s; it remained virtually unsettled and home to many buccaneers and pirates until the Dutch, with buccaneer protection, established a permanent settlement in Tortola in 1648. The French captured the islands in 1668, and the British took them away in 1672 and started colonizing. They were given different colony statuses in 1872. Today the British Islands are a Crown Colony with its own government. (Don't ask us why they used US dollars there!) The pirates were very happy with all the islands' hiding places to stow their loot from the ships they had knocked off from the main shipping routes. It was a surreal experience to sail quietly past these beautiful islands imagining the sins and adventures of a time long past...rather like a dream.

Back to our adventure. Next stop, North Sound, obviously on the northern end of Virgin Gorda. "There are numerous world-class restaurants and marina complexes here to suit everyone," boasted the cruising guide. Hmm...were! Blasted raw by the hurricanes, only Leverick Bay Hotel remained operational. The others were all in the very early rebuilding processes. Some of them were large establishments. So sad, but that being said, even with the coral being tossed, turned, and heaved up on beaches, the whole environment was still so stunning. The water was a deep sapphire blue, clear as a bell for at least twenty metres or so deep. Brilliant white sands fringed by seagrass attracted the turtles, which constantly bobbed up beside us in the water. As we headed towards North Sound, the highlight for me was going past Necker Island, the private home of Sir Richard Branson, owner of the Virgin empire. You could stay there if you were so inclined. Prices started at a cool US$80,000 per night. It had been wiped out by the hurricanes as well, but clearly, Sir Richard had a head start on his renovations.

Since arriving in Caribbean waters, we had not yet been over the side for a swim; it looked irresistible and didn't take long before we were all in the water. Much to our surprise, we discovered the hull of *Marmax* was covered with a thick, swaying carpet of alien-like creatures.

These creepy things were gooseneck barnacles. They apparently attach themselves to almost anything floating and are found worldwide in warm, tropical waters. In some countries, they even eat them! Out with the scrapers, we spent hours cleaning that hull! The growth was incredible. How could they possibly cling to the bottom of the boat with the force of the ocean continually buffeting the surface? Suckerfish, like zebra-striped barbarians, nibbled on the remains and swam about us, waiting for scraps. They were to become our friends and allies in the next few months as they methodically vacuumed our hull as we travelled home through the tropics.

Another exciting encounter was the sight of this enormous private yacht called *A*. It could be called many descriptive words, but the nice ones escaped us. It was apparently worth US$450M, the largest and tallest private sail-assisted motor yacht in the world and owned by a Russian billionaire. It had fifty-four crew, was 143 metres long, 25 metres wide, and eight decks high with a private submarine and an underwater observatory. Google it; it's fascinating! Its tenders looked like flying saucers running around the bays but shaped like cigars. These fancy zoom-zoom tenders were zapping everywhere; we always had our eyes on them when swimming or snorkelling. There did not appear to be any speed limits or rules, even in the middle of commercial harbours. We read a story written by a lady sailor who recently had her head split in half by the props of such a boat; she narrowly missed death. This speeding tender accidentally ran clean over the top of her and her husband in their own dinghy. The zoom-zooms are so big and move so alarmingly fast!

After exploring the devastated beachfront establishments, we motored across the harbour to Leverick Bay: picture-perfect and utterly divine. Plenty of money had been poured in here to get this little tourist hub up and running after the onslaught of the devasting hurricanes. The only complaint was the cocktail glasses at the bar were plastic. Gee, it was expensive! Les, ever the fashionista aboard Marmax, indulged in buying his BVI cruising attire while here. He was like a kid in a candy shop. I swear he is a half girl! Truly a millionaire's paradise.

After a few days, it was now time for us to start thinking seriously about making our way south down the chain of islands that make up the Caribbean. We were well aware the hurricane season was fast approaching and could see that many boats were starting to make haste to move out of the area. However, we were in dire need of a rigger, as we still had two of our main reefing lines to be replaced before the long journey ahead. We also had to be out of the Caribbean and into Panama by June 1 for insurance purposes. Simple, you may say, but not so simple.

We headed to the main island of Tortola, named by Columbus in 1493, meaning turtle dove in Spanish. Also, the only place you could find a rigger. Anchoring for the night in a little place called Buck Island, we headed around to Nanny Cay Marina the following day. We were going to attempt to find the cousin of our Kiwi friend Carolyn. Her cousin Jane, another fellow Kiwi, managed the hotel and accommodation there. Sometimes it's not what you know but who you know. We tracked the delightful Jane down and asked lots of questions. What a fantastic find she was in the short time we had with her. Thanks, Carolyn and Jane!

The riggers of Nanny Cay solemnly shook their heads; they could not start another job for at least a week. This was a bit of a blow, as time was beginning to run out for us. We headed up the harbour to Road Town, the capital city of the BVIs.

This was a commercial harbour. It was daunting trying to figure out where to safely anchor Marmax. There were fishing boat moorings fouling the seafloor everywhere, big tourist boats with giant bow waves steaming past, and foreign yachts sprinkled all over the place. We nestled in amongst this soup of vessels and chaos, gingerly dropped the anchor, locked the boat securely and went off to explore the town.

They must have been fierce hurricanes. Wow, what a mess! Complete destruction was evident everywhere, and businesses were trying to sprout out of the debris like green shoots in a rock garden. How strong must a wind be to punch holes in concrete buildings? Shopping centre facades were simply pounded to pieces, empty windows with gentle breezes wafted through them. The uneven pathways were still full of gaping holes, the sewerage systems were still not working. Broken trees, hardly any birds or animals, just chickens. But still, the people were happy, the traffic was hectic, and half the cars looked like a coconut tree had fallen on top of them. Gaffer tape was clearly a necessity here.

Beaming smiles radiated from the local Rastafarian women who ran the colourful craft markets on the waterfront; they had just opened their shops for two cruise ships. Their thickly braided hair was fashioned into perfect cornrows, black and glossy. We felt sorry for them in their struggle to make a dollar while coping with the devastation surrounding them.

Taking a stroll through the back streets revealed all sorts of architectural delights. Beautiful old West Indian homes with new, bright red and blue roofs. Lots of Victorian dado work all around the porches. Red and white churches, Her Majesty's Prison, and commercial buildings in such brilliant colours always put a smile on your face. Buildings were perched atop huge boulders on a hillside.

It was all rather beautiful and a feast for our eyes. Still, it was rather alarming to be reminded that the hurricanes had struck so long ago – two years before! I could not find the local paint shop; I wanted to hear the sales spiel they would give their clients while deciding on paint colours.

Thankfully, we eventually hunted down some Road Town riggers. We got the same story; they could work on the rigging "next week", so come back. Bearing in mind we were on "Caribbean time", I was personally dubious about this "next week" business. We could not afford to spend time on marinas just waiting for these guys. None of their phones appeared to work too well, or did they not answer? You physically had to chase them in suffocating heat. It was not fun!

What was fun was the famous Pusser's Pub, renowned for their Pusser Painkiller cocktails! You've got to love it when you order one. "Would you like two shots, three shots, or four shots, honey?" Oooooo…! Wouldn't I love a four????? But common sense prevailed, and we had a long way to walk home with heavy loads. Oh Lord, they were delicious!

We had tied our tender to a jetty earlier on and connected to another pub, called "The Pub". Otherwise known to us as the pub with excellent wi-fi! It was time to catch up with communications. At the same time, we waited for a man to deliver three LPG gas bottles, which would hopefully last us until we sailed back to New Zealand in October.

We had a few days to kill before the riggers would hopefully materialise, so we continued our extensive exploration of the BVIs. Jane had mentioned her favourite spot was in Little Harbour, Peter Island, where you could swim with the turtles, and indeed we did! What a thrill for young Janelle!

The next day, we went snorkelling around the caves of Treasure Point at Norman Island. The water was so unbelievably clear it hardly seemed there. This was the island of legends about pirate treasures. Tropical fish swam all around us for hours. You seriously just did not want to get out of the water, but it was hungry and thirsty work playing "fish" on a hot day with masks and fins on. By now, we had sourced the recipe for those "Painkillers", we also had the ingredients. What better way to end a day in paradise? Cheers!

The Caribbean is obviously famous for its multitude of enchanting bays and islands. We were lucky enough to have time to either explore or circumnavigate almost all of them. There were a few fantastic standouts, starting with Dead Chest Island. Allegedly, this is where Blackbeard marooned fifteen of his crew with nothing but their sea chests and a bottle of rum. Hence the line from that old favourite, "Fifteen men on a dead man's chest, Yo, ho, ho and a bottle of rum!"

Leverick Bay

Sunday, April 28, 2019

8. The BVIs – Never A Dull Moment!

As part of our immigration fees, we purchased a National Park entrance card when we arrived in Virgin Gorda. This was to give us access to enter what is known as the Baths. We had been advised to get there at 7 am to beat the masses. No crowds here, hallelujah! As it turned out, we had the whole place to ourselves for most of the time.

The Baths and Devils Bay National Park on the southwestern tip of Virgin Gorda was a magic place resembling a scene from a sci-fi movie. Massive granite boulders were strewn up and down the beaches, piled on top of each other. They date back to some seventy million years ago. It is fascinating that many geologists believe they were formed by lava flowing through cracks in the ocean floor. Over millions of years, the seafloor rose, and the exposed boulders were sculpted by wind and waves until they formed their current shapes. Another lot of geologists argued they may have been brought down the Caribbean within the glaciers of the last ice age. When the ice receded, they left the granite boulders behind. Why was this the only place in all of the Caribbean where these formations were found? They are one of the wonders of the world in our eyes!

Enormous round rocks formed a labyrinth of caves and passages. Sunlight filtered through the cracks revealing shimmering, clear saltwater grottos and warm pools. It was just fabulous to lay around in and incredibly not at all ruined by tourists and their many feet! We had an absolute ball in here; it's too hard to describe how special it was...

After exploring the National Park, Bruce and Janelle took the tender back to the boat. Glenys, Les and I took the long dusty road around three bays to check out the stunning holiday homes built on top of the boulders hugging the picturesque coastline. We chatted with a few random people along the way. We were trying to find our way along the shoreline towards *Marmax*. We had discovered a narrow opening in a rock formation to crawl through, spilling out onto the sand of what appeared to be a private beach. Three men were sunning themselves in the searing heat. One very plump and sunburnt man was floating about in the water in a yellow, plastic blow-up banana. After a friendly chat, we were surprised to hear that this was his twentieth trip to this particular holiday home in Virgin Gorda. They were American businessmen. After a while, the man in the floating banana cocked his head thoughtfully, asking us where the heck we had appeared from.

"Um, out of the rock?' replied Glenys, trying to hold back laughter bubbling up inside her. Not sure if it was the look of him floundering about on that banana or a nervous reaction to us finding ourselves on a private and secure beach, clearly trespassing.

"Isn't it dangerous?" he asked.

"Nope!" Glenys was losing it. She had tears in her eyes, and a deeply suppressed giggle was about to explode; it was time to get out of there.

All these years spent holidaying here, and the guy had no idea there was an entrance through that rock leading to other beaches.

Dressed in our elegant resort wear, Glenys in a sarong, me in a short sundress, and Les in his usual T-shirt and shorts, we walked into the water and kept going. We started swimming out to *Marmax*. She was on anchor, hidden behind the boulders and well out of sight of the Americans. I turned my head to look back and saw their looks of amusement. It kind of felt like we were in a 007 movie, though Glenys and I did not quite look as alluring as Bo Derek! They must have thought we were a bunch of total nutbags. I have often wondered what those three American businessmen thought of us as we disappeared into the ocean.

During our explorations of the bays, we came across a place called Marina Cay. Although the wind was blowing its socks off when we arrived, we simply could not go past this little damaged gem without supporting them in some way. It was a tiny cay with quite a lovely story behind it. In 1937, Robb and Rodie White bought the six-acre island for US$60. It was a lonely little bit of land called Diddledoe Island. Robb was a writer, and with his beloved wife, Rodie, they ran off for peace and inspiration. They built a cottage on the hill and left in 1940, as Robb had to go to war and Rodie had to recover from an attack of appendicitis. While they were gone, they sadly lost the title to the island and never returned. Robb wrote a book about their three years on the island. It was titled *Two on the Isle*, which for those of you who are movie buffs, later became a movie in 1958 starring Sydney Poitier and John Cassavetes. How did we support them? In a makeshift bar and restaurant hanging out over the beach, we partook in several piña coladas, of course! Everything else was shut down and unfortunately blown to pieces, by yes, you guessed it those two terrible hurricanes.

The next discovery was appropriately named Lee Bay because it provided excellent shelter in the lee of the hills from most winds. This was a fascinating, in-the-middle-of-nowhere place where we met a Kiwi guy, David, who was on a two-year cruise with his son Zack. That night we enjoyed drinks aboard *Marmax* and a Kiwi pikelet tea the following morning (compliments of Janelle) aboard their gorgeous 54-foot yacht. What a spot! Fish were leaping out of the water everywhere along the rocky shore. Grey pelicans, big and clumsy, were smashing into the water, eating the fish. They were also teaching their young to fly and land on the water. Black-headed gulls jumped about on the backs of the pelicans, trying to steal the fish they had just caught. This was a performance that went on all day long, just awesome to watch.

Then finally, Cane Garden Bay, around the back of Tortola. Arguably known as one of the prettiest and best loved of all the BVI anchorages. Some say Jimmy Buffet's songs "Mañana" and "Tire Swing" helped in its popularity. This place was so much greener and lusher than the other sides of the island. Though suffering damage, much of the beautiful West Indian architecture had remained intact. Two Painkiller drinks for the price of one at the Paradise Bar! Well, it was there; we wiled away the night as the sun softly set over the boats in the bay, and the grooving sounds of Caribbean reggae helped the drinks go down smoothly. What a heavenly life!

It was time to get back to Road Town, find those elusive riggers, and stock up the boat again. The following planned route was back north to Anegada, the Drowned Island, to check out flamingos nesting. We then hoped to head south, down the eastern chain of

Caribbean islands en route to the Panama Canal, with an estimated time of arrival over there around May 30.

We stayed in the Village Cay Marina in Road Town for two nights, mainly for the riggers to sort out our reefing lines through the boom. We also needed to do our final big shop for stores, especially cans of food and a few months' worth of alcohol. The supermarkets were so frustrating, as only half of their goods had price labels. We, like all the other travelling yachties, were constantly having to ask for price checks. It was bizarre; the prices varied so much from island to island and shop to shop. A can of coconut cream fluctuated between US$1.50 to US$4.75. Add the day's exchange rate at sixty-six cents to the NZ dollar; it all added up. Mangoes had a variance of up to US$4 each. Shopping took hours! So with trolleys full, we headed back to the boat. We had to leave a $50 refundable deposit for the trolley cart. Glenys and I found a lovely big employee of the store to escort us back to *Marmax*. No $50 was needed.

The next day, I had Les with me to get his load of beer. The $50 deposit had to be paid. Bang, crash, through giant potholes, rods of construction steel hanging out of the concrete roads, down dusty embankments, to the marina, which was also restored after the hurricanes. Ker-clunk, ker-clunk, ker-clunk! No wonder all the grocery trolley wheels were in various stages of collapse!

The Village Cay Marina was right in the heart of the action. A quaint waterfront restaurant and bar hung over the clear water. There was also a marina office with a yachties' book swap, a serviced laundry, and full shower facilities for us travellers; these were just what we needed. Glenys and I had started looking a bit unkempt in the hairdo department. I hate grey hair; the salt and pepper look was beginning to overrun my head. It was time for action.

I made no crew announcement or anything, so no one knew I would dye my hair that morning. I went up to the bathrooms; they were clean and very roomy. I thought I would throw a permanent colour in my hair, wait for twenty minutes while it cured, rinse out, and voila! I would be back to my usual, sparkling self. I mixed up my potion, a two-pack for guys who don't know our womanly secrets. I very carefully painted it in with an artist's brush, section by section. Tick, tock…twenty minutes curing time was up. I went and turned on the tap. No water came out; this shower seemed out of order. I packed up and moved all my gear into the next stall; there were only two. I again turned on the tap, ready for a steady stream of fresh warm water. Zero, zilch, silence. Oh boy, what was I to do? The outside basins were too small to stick my big head under the taps to rinse the dark brown dye. To be honest, I even looked down the toilets to see if they would be deep enough to swish my head about it.

I stared at the little notice on the wall. "Do not put any paper down the toilet" So where exactly do you put it? One look at the open bin on the floor immediately put me off that idea. I was getting frantic; I was looking an absolute fright. What if someone came in? Frustration crinkled my eyes as I looked up at Spanish writing on a green slot box; did I need money? I hurriedly emptied my little travelling purse onto the bench and shoved some pennies in. No joy, they all fell through to fall at my feet. What was I doing wrong? To get back to the boat, I would be forced to walk through the restaurant looking like a lost wild witch. Where were all the women who usually visited bathrooms? I had no visitors. There was no choice but to tiptoe incognito along to the marina office. Turns out that I needed marina tokens to work the showers; why had I not been told? Fifteen minutes later, I sashayed my way down to the yacht, feeling a million dollars. Another lesson learnt for me!

While at the marina, Les had gone for a poke around the yachties' book swap in the marina office. Low and behold, he came across a beautiful Merriam-Webster Dictionary and Thesaurus. "Shhh…" Les whispered to me. "Don't tell Janelle about it; I'll give it to her as a present when we are back out at sea!"

Glenys went for a wander in the kayak the next day for an exploratory paddle around the marina basin. There w e re d o zens and dozens of smashed up boats still lying around the harbour and mangroves, many sunk with masts sticking out of the water. Such a sad, sad sight. She came across an Australian lady aboard a yacht on a mooring. This was the lady's second trip t o the BVIs. She was about to take their second yacht back to Oz to sell after restoring the hurricane damage. Insurance companies had been selling damaged boats cheap. The local dive shop owner was well on his way to becoming a millionaire salvaging all the stainless-steel fittings, rigging, winches, sails, anything sellable and salvageable. He had done some deal with the insurers. Good luck to him; it would not have been easy money. The stainless-steel flybridges, pulpits, etc. tangled up in the mangroves alone would make you cry when you think of what the stuff is worth in other countries around the world.

Les was adamant he would go to the movies; he is a movie buff, and the rest of us are not. The new *Avengers* was showing in town, and he was not going to miss it. Off Les trudged alone, forgetting I was holding all of our cash. After walking for fifteen minutes in the oppressive tropical heat to get to the theatre, he realised he had no money. They would not accept his V isa card, so he marched fi fteen minutes back to one of the few cash machines in the town. A fifteen-minute walk returned to the movie house, soaked in sweat by this stage, but just in time to enjoy what was apparently a brilliant movie. The big bonus? It sounded like the best building in town; it was air-conditioned, even though it looked like it had been machine-gunned down by a military tanker from the outside. Les was a happy man!

Bruce, however, was, let's say, thunderous at having just received the bill for the three riggers who had worked intermittently for two days on *Marmax*. Though happy to have the job done and done well, the blood from his face reflected what had just left his bank balance. We headed off for his commiseration drinks at the famous Pusser's Pub. There, we met up with Jane from Nanny Cay again. We caught up with all the local gossip, imparting more of her valuable knowledge of the area we were about to travel. (Glenys and I had no idea that all these beautiful island-born Caribbean ladies here wore wigs and hairpieces!!) Les was astonished to find his beloved Australian Rules Football playing on the television at the pub. Here he was, thousands of miles from home, and in the middle of nowhere, he could watch, what he believed, was the greatest game in the world. A true Aussie! We downed an incredible meal, and Bruce felt a whole lot better as we walked home with a crooked trail in the moonlight, his keepsake, an enamel Pusser's Pub cup, clasped in his hand.

The following day, with the wind still steady at 20 knots, we headed off to Anegada. Moments before throwing the lines off the marina to lead off to sea, Janelle sidled up to me in the cockpit. She sighed with relief and, looking coyly up at me, revealed she had gone into the marina office with Bruce to complete paperwork when we had arrived in Tortola. In the office, there was a book swap with a stinky dictionary on the shelf. "Well," she says, "thank God no one found it and now we are leaving."

The little monkey! We had no dictionary or thesaurus on board for her classes, which as you can imagine, had become a significant handicap for her learning for Year 10. Remember, there was no wi-fi, so no Google aboard. We had struggled to get this girl interested in schoolwork at the start of the trip; she had many higher priorities in her mind. Bruce had suggested downloading the dictionary earlier on in the journey. Little did Janelle know that Uncle Les was one step ahead of her and had already swooped on it in the marina office. The delight on Les's face when he presented that dictionary to her at sea was just gold! Janelle was crushed while he stomped around like a victorious Rumpelstiltskin, all of us howling with laughter. The Webster proved very popular and was consistently used, much to Janelle's disgust. High five, Les!

Entering the bay of Anegada, we realised that *Marmax* would draw too much water to get into the moorings in the harbour. We resigned to anchoring off one of the long sandy beaches, probably around eight hundred metres off it, as it was so shallow. With the wind still building, Bruce wanted us back to the boat within the hour. It was an impossible ask as it would have taken us at least two hours to get ashore and find the flamingos. Consequently, we had to flag the whole idea and sail back to North Cove, hard on the wind once more.

I was thrilled to sail close to Sir Richard's Necker Island on our return, so we were able to examine it in greater detail through binoculars. Man, what a pad! A sandy cay lay within walking distance of his island. It shone like a precious, gleaming jewel. Three evenly placed, matching palm trees appeared from a distance as if they were growing out of the water. Perhaps this was his private drinking island, or is this where he interviewed his air hostesses? It was all very dreamlike.

If Richard had been sunning himself on the deck of his grand home, he would have heard our screams of delight at our next sighting. A trio of stunningly fluorescent orange flamingos appeared from over the waves. They flew past the boat, only a foot or two off the surface of the water. We had never realised they were such a bright, shocking orange in colour! It was like seeing three brand new flying life jackets come zooming over a bowl of wobbly blue jelly. It was a remarkable sight; we felt so privileged to have seen them!

After a night back in North Sound for shelter, we moved to Virgin Gorda for customs clearance. We had initially planned to clear port in Jost Van Dyke, but we skipped it to head south with tropical weather starting to build. We were in and out of Virgin Gorda within two hours. A far cry from the last time we were in there. The marina was close to empty. It appeared we were among the last of the travelling yachts to head away from the oncoming hurricane season. Once we had fuelled up, Bruce cleared customs. After farewells to Angela, our Easter security guard and one of Janelle's new little friends, we were off.

A final salute to the magical coastline of southern Virgin Gorda, the Baths, and Devils Bay Beach. We blew our kisses and settled into a steady slam, again, on the nose while sails were reefed down. Carefully, we sailed through the choppy channel between curiously named Fallen Jerusalem Island and Virgin Gorda. The night continued with lots of solemn backward glances to watch that beautiful bunch of islands fade into the sunset. It sure had been the trip of a lifetime, and the British Virgin Islands would never be forgotten.

We were not dreaming; we had just been to paradise!

Road Town—Tortola

Janelle at "The Baths" - Virgin Gorda

The Riggers finally at work

Garden Bay—Tortola

PUSSER'S PAINKILLER

Good for everything and probably the smoothest drink you'll ever taste! A delicious blend of Pusser's Rum, cream of coconut, pineapple & orange juice served over the rocks so the ice isn't blended into the drink to dilute its delicious flavor. Depending on the severity of the pain being experienced, you have your choice of Numbers 2, 3 or 4 designating the relative amount of Pusser's Rum! Be careful, this is a smooth and sneaky drink and a giant 18 ounces!

Friday, April 26, 2019

9. We Be Cruisin' the Caribbean!

Around this time, we had just received news of a dreadful pirate attack on a Kiwi sailor, Alan Culverwell, and his family in the San Blas Islands of Panama. After hearing of his brutal murder, if we weren't all sleeping with an ear out in port for intruders before, we all were now! There had also been another attempted attack on a 55-foot Bénéteau yacht down south only a couple of weeks back. These guys had been in heavy seas. Fortunately, they were able to tack (zigzag) their way away from the eight armed pirates, not without copping gunshot damage to their hull.

Consequently, our plan was to skip the lower Caribbean and head out from Antigua across the upper latitudes. The goal was to sail just below Puerto Rico and Jamaica en route to Panama, rather than the lower Saint Lucia, San Blas Islands path. The Yacht Services Association was working with the Coast Guard, stepping up protection for all travelling yachts. Life is so cheap in some of these countries, it was not worth the risk.

Saint Martin (French)/Sint Maarten (Dutch) was our next port of call. We entered the gorgeous Simpson Bay and dropped the anchor outside the harbour with around thirty other boats of various shapes and sizes hanging off moorings. Seriously, the water there was so blue, it looked like copper sulphate. You could almost be worried it would stain your hull blue; the intensity was so incredible. I, for one, could not stop staring at it. I wanted to take a thousand photos; actually, I think I did. I was completely head over heels in love with this stuff!

This place had quite a history, going back over five thousand years under several occupations. In 1648, the French and Dutch signed a partition treaty taking reign from the Spanish. The story goes that, to divide the island, each party had to choose a walker. Positioned back to back, each man had to walk, not run, in the opposite direction following the shoreline then meet somewhere in the middle to divide the island. The French were left with fifty-four square km, and the Dutch thirty-two square km. Some say the French chose wine as a stimulant, whereas the Dutch chose Dutch Gin to quench their thirst; hence the greater difference in the area claimed. Many also say the French ran instead of walking. Unbelievably, in the next few centuries, Saint Martin changed nationality eighteen times!

To enter the marina basin, vessels must go under the Simpson Bay drawbridge. This bridge only went up three times a day: 10 am, 3 pm, and 5 pm daily; the times were strictly enforced. There was no waiting around. This was also one of the main airport roads that serviced the Princess Juliana International Airport. You know, that famous beach you can stand on and get knocked over by the jet stream of planes taking off or landing.

Bruce and I leapt into the tender to go clear customs ashore. We eventually found customs at around 9.50 am. There were no signs on buildings or anything. You just had to find your way through the rubble until you stumbled upon it. This was located in a very non-descript, hurricane-damaged structure right next to the drawbridge. Customs clearance was generally a long, tedious task with many questions and endless ridiculous repetitive paperwork for Bruce. Lots of useless mumbling and fumbling of paperwork between office officials who clearly have all the time in the world and no pens that ever work.

Fingers drumming in the oppressive heat, I muttered, "We're not going to make the ten am bridge rise, Bruce." Another wasted day? At 10.04 am, we fled out the door – it was a super speedy clearance! Much to our delight, Les, Glenys, and Janelle had seen the drawbridge rising and hurriedly pulled up the anchor. Flags aflutter, *Marmax* came proudly steaming up the channel, under the lifted drawbridge, joining the procession of other foreign yachts going through.

This bridge also marked the border between the French and Dutch territories. The marina basins were full of hurricane debris; some of them looked like a kid had emptied his toybox upside down. Broken boats and shattered marinas, yet the St Maarten Heineken Regatta celebrations, part of the 2019 Caribbean Regatta season, had been held there in March. A massive effort had taken place to get it all partially cleared up for this flamboyant annual event.

We changed course several times to weave our way through sunken vessels and mangled marinas. We endured another hairy arrival in 20 knots, reversing into the berth. Damn, we were getting good at this! It seemed so easy, writing that sentence, but in reality, it was not! Two boats were alongside us with really nasty blue concrete gashes on their sterns; we were determined not to damage our *Marmax*! The wind did not make things easy, but we landed safely. We were very excited to be there.

Two nights were enjoyed by us in this vibrant and colourful port. It was the first marina we had come across geared up for worldwide travelling yachts. The BVIs had been very much focussed on the charter yacht crowd, especially for those big catamarans. There were not many boats in port, but the few there were gearing up for very long trips. The modern and mighty 78-metre yacht *Venus* was in port. This beauty was built for entrepreneur Steve Jobs of Apple fame.

There was plenty of short-range wi-fi outside the gates of the marina development, so we drifted around garden edges and the waterfront to lock onto it. Jimbo's bar was right next door, but you couldn't drink all the time…could we??

St Maarten was quite amusing. There was plenty of blown-out shacks and hurricane debris still about. We discovered the most creatively designed shipping container buildings we had ever seen. There were upmarket casinos, empty shopping centres, big flashing electronic advertising boards, exclusive upmarket duty-free shops, modern apartments and houses, dusty potholed roads, and more foolhardy drivers weaving madly amongst mopeds and helmetless bicycle riders. Sprinkled amongst all of this were quaint, busy street stalls and West Indian shops booming in riotous colour and Caribbean music and laughter like hundreds and thousands being shaken over it all. We devoured a plate of delicious French pastry delicacies in a nearby bakery, all the while kept amused by huge frilled guanas darting about our feet. These laid-back lizards are fascinating and can outstare you, hands down. They also scarper quickly if frightened; the place was full of them. There were loads of traditional art, food, and clothing and a vast range of cuisine from French bakeries to Chinese restaurants, clique coffee boutiques, and open West Indian rib shacks. Just a feast for your eyes no matter which way you turned.

We enjoyed a rum tasting session that night with Mr Topper, a prominent rum bar owner. At the sprightly age of eighty-one years, he had children aged from sixty down to seven years of age. That man still had a significant twinkle in his eye, or was that the reflection of the gold and diamonds dripping off him? A merry night was had, with the smoothest rum in the world, Toppers Rhum – keep an eye out for it! Sold in thirty-six countries and coming to a town near you!

Glenys discovered it would cost only US$1.50 each to catch the local taxi bus to the Dutch-side capital of St Maarten, Philipsburg. Due to the prevailing winds, it was easier to bus rather than sail there. This was where massive cruise liners were offloaded. Loads of bars, duty-free shops, boutique clothing, restaurants, and street markets all over the show. Guys were holding out cold bottles of Heineken beers to us at some of the shop fronts, offering a free beer if we came and looked at their shops.

Glenys was lured into a ritzy duty-free jewellery shop with some of the most exquisite pieces I had ever seen. To cut yet another long story short, Bruce bought Glenys a stunning diamond eternity ring, much to Glenys's delight. The shop owner offered them both a glass of champagne to celebrate. That was probably going a little too far for our Bruce. Being the good sister that I am, Glenys and I partook in a glass each while a butterfly flew away with the remains of Bruce's wallet.

A fine time was had exploring this beautiful part of the world, which beyond the main waterfront precinct, really had not lost its Caribbean magic at all. Some very annoying street hawkers were bleating for our custom; they hammered those passengers arriving from ships! So many of these cruising ports were woefully destroyed by consumerism; St Maarten was a delightful surprise. We always felt safe, and the people were all friendly and helpful – then we took the bus to the French side! Hmmm…we spent very little time there.

Our taxi-bus deposited us into the township of Marigot. The part we saw was still very run down. There appeared to have been a minimal effort in the hurricane clean-up compared to the Dutch side. Filthy, drug-induced street people about, sad-looking buildings, plenty of armed military walking the streets. It was, surprisingly, the opposite of the Dutch side. No smiles, no happiness, it just seemed… well, a sad place with not much to offer. With a 14-year-old girl in tow, this was no place for mum to let her out of her sight!

Back on the bus, we were, by now, foot weary and ready for a swim in the pool at the Simpson Bay Yacht Club. Unfortunately, the season was already in shut down; the bar was closed, so we lolled around in that glorious empty pool alone, caught up on the wi-fi, and strolled back to *Marmax*.

The following day, after clearing customs, we left the marina and headed to the drawbridge. It was like the start of a yacht race, as we had to be lined up fifteen minutes ahead of the marina basin bridge going up. 5, 4, 3, 2, 1…the drawbridge went up, and off we all surged, only to have a false start! An ambulance came screaming along the road with its siren blaring. The bridge went down; our throttles got thrown into reverse. Some sizable yachts were trying to maintain position amongst hundreds of mooring buoys in a frisky wind. The drawbridge went up again, and we were off. Goodbye, beautiful people! We hoisted sail as the aeroplanes came roaring to land alongside us at the Princess Juliana International Airport. Next stop, St Barthélemy, only a 20-nautical-mile sail away.

Our anchor was dropped in a stunning volcanic bay for the night, very rugged with a crystal clear seafloor. This was a nature reserve with numerous turtles spotted throughout our stay. Early next morning, Glenys, Janelle, and I took the tender ashore to go running around the hills and rocks to see what was on the other side of the island. We yelled from the highest rocks we could climb. Wild goats darted out left, right, and centre. Smashed up prickly pear trees and cacti smothered the hills. If we had stumbled and fallen, we would certainly not have wanted to grab the nearest tree or stick! Glenys thought the sharp, prickly branches would be fairly handy as pirate deterrents around the outside of the rails of *Marmax*.

There were glorious panoramic views from high up in the rocks; we were, to be honest, dangerously high up there. We could see Bruce and Les mucking around with our storm sail, which had never been unpacked to this point. Well, you would not miss this little "black duck" in a storm! The sail was fluorescent orange; that was different!

The main port of Gustavia was given a miss by us, mainly because A) we did not need to enter it, B) we were warned it was expensive, and C) to be honest, it didn't look that inviting or exciting from the sea. We moved down the coast to another sapphire blue cove to shelter from the wind, do some snorkelling, and kill a few hours before heading on with an overnight sail to Antigua. If we left earlier, we would have arrived in Antigua in the dark.

Apart from Les catching a prize mahi mahi at sunset, the night ran smoothly with a steady 15 knots of wind, again, on the nose. It was surreal riding along in the clear, cloudless night through the leeward islands of the Caribbean. The tiny islands and various countries looked like cruise ships dazzling on the horizon, all lit up and sparkly as we cruised on by. Antigua would be our last stop off, as the tropical rains were building daily. You could simply feel it was time to get moving, with the air getting heavy with moisture. The final remaining yachts that had been alongside us in the marina were all moving out. Even the marina bars, restaurants, and shops were closing up for three months from the following week.

We arrived in the famous Falmouth Harbour. *Marmax* berthed in the prettiest marina imaginable, right outside the Antigua Yacht Club, with everything within a short stroll. The priority was to get our gas going so that we could cook! It seemed the solenoid was the problem and had packed it in a few days prior. It was Saturday, and there was not much help around, although some great chandlery shops were about. We were lucky to have a power generator aboard, plus an electric jug, toaster, microwave oven, and a shallow paella frying pan. All was not lost; it just limited us with what we could cook. We had at least eight to ten days at sea ahead of us, approximately 1300 nautical miles to Panama. I was pretty sure the crew would get sick and tired of porridge and rice if we did not get the gas going. Thankfully, Bruce managed to get it fixed. We were cooking with gas!

The next day we caught the local bus to Saint John's, not only to get more spare gas fittings but to check out the heart of town and walk amongst the real locals. What an experience! So much vibrancy, street music, the sweet smell of tropical fruit stalls and pastries baking. The town was an absolute madhouse; the islanders were out in full force getting their weekly supplies. I couldn't take too many photos, as we stuck out like a sore thumb. We guessed taking their picture may have made them uncomfortable, even if we had asked.

Two big eye-openers for us! The fish markets were full of tiny coral fish with flies buzzing about; it was hard to believe the locals also ate pufferfish! And two large yacht transport freighters, tied at the main wharves, loaded to the gills with vessels being shipped back to the USA. These were yachts that had completed their seasonal pilgrimage to the Caribbean. Many seemed hurriedly packed, as they still had their sails on their booms and flags flying from their stays. They were part of the fleet that had just competed in the St Maarten Heineken Regatta and Antigua Race Week. Quite a fascinating sight for our crew; we had never seen these transporters before.

St John was more authentic than most other towns we had visited so far. It was full of smiling, colourful people enjoying life. The sound of women giggling and the men doing their fist-bump greetings was a constant. We brunched at one of Ernest Hemingway's old haunts. After losing Glenys several times, we managed to find our way back to the bus depot and took the perilous ride home. There were no speed limits, only speed bumps, and I swear there were no shock absorbers, indicators, or brakes in that vehicle!

Time to leave these fair shores. We battened down the hatches, which included the tender being deflated and outboard stowed in the stern. The Antigua direct to Panama course would be our safest route to stay well away from the likes of Venezuela and Colombia. Allegedly, pirates around here had a notorious appetite for outboard engines. It was a shame we had to miss the San Blas Islands of Panama; apparently, it was a real-life paradise. The recent murder of the Kiwi guy was on our minds. It was not worth the risk to our family at that particular time. *Marmax*'s fridge, cupboards, and lockers were repacked and made secure. All safety gear was checked in preparation for our long journey ahead.

Marmax enters the Simpson Bay Draw Bridge

Cane Garden Bay

Norman Island

St Maartens

Hurricane damage everywhere!

10. A Smooth, Safe Ride: Antigua to Panama

I t was with heavy hearts that we threw the lines off from the Antigua Yacht Club marina. We anchored out in the harbour for a couple of hours, preparing *Marmax* for her next leg across the Caribbean Sea. With sails hoisted, we drifted about on the anchor while the three new reefing lines were marked off on the winches and general maintenance was carried out. The spray dodger was taped up again, as it had weathered a fair few ocean waves since Glenys and I had last strapped her up in Gibraltar.

Leaving at 1.15 pm, we had a crisp southeasterly breeze, and we headed downwind in a pleasant 10–15 knots on a beautiful blue ocean. One of the Caribbean's most dramatic islands, Montseratt, was in full view. Known for its soaring peaks and rainforest-covered hillsides, we could see black flows of lava that had frozen in time. The Soufrière Hills volcano had erupted in 1997, killing nineteen people and devastating the south of the island; it had forced the removal of the entire population to the north. She was quite a stunner from the sea.

I was facing the stern, randomly chatting away, when in full vision, a massive burst of white foam and spray shot up around thirty feet in the air. Instant panic!!! We had not passed that close to a reef!

"What the heck was that?!" I gasped.

The next minute, this enormous humpback whale came firing directly up into the air with its pectoral fins twisting sideways. With a mighty whoosh, it came crashing down. Wow! All hands were on our three sets of binoculars. More screams of delight! On an otherwise uneventful long-distance sail, the thrill of seeing such an incredible performance of nature got the crew jumping around like a bunch of school kids watching fireworks! At least three of these magnificent whales were performing, shooting up, then crashing with water in all directions. I wished they were ahead of us, rather than behind, as we had to sail away from the show. Sadly, we did not get photos! It was a pretty exciting welcome to the vast voyage we were beginning. Thank you, universe!

We were not nearly as apprehensive at the start of this crossing as when we started the trip across the Atlantic. It can be a rocky road living in close quarters with family members for months on end. During long ocean voyages, relationships can really be tested, whether they are family or not. We all had different personalities, but we'd worked each other out by now. We got along most of the time, but most importantly, we had total trust in each other's abilities. Firm round-the-clock routines were established, and we were also very confident sailing *Marmax* in variable weather and ocean conditions. She was still a relatively new acquisition to us; we were still getting to know her.

The Caribbean Sea was very flat compared to the Atlantic, and there were many more birds about. There were no dolphins but a constant pageant of small flying fish bursting out of the waterlines and across the waves. All day long, they skimmed along the surfaces in directional schools. Quite mesmerising! It looked as if a giant was throwing handfuls of diamonds across the surface of the sea.

Bruce had a birthday aboard, a cause for celebration. By now, we were into the seventh day of our crossing and by far the most tedious yet. The weather had been so perfect to date, 10–22 knots downwind. Still, no dolphins could be seen anywhere. There were a few passing ships at night and, unfortunately, more miles of that pesky Sargassum seaweed streaming alongside us. This was the bane of Les's life, as we were constantly egging him on to catch fish for us, but the weed was just so thick. If it got caught on his lures, he had to waste time and energy continually hauling the line into the stern of the boat. If we were sailing at a speed of more than 5.5 knots, we were simply moving too quickly to fish. Once we reached a hull speed of 7 to 8 knots, the rod was decommissioned and strapped to the mast in the saloon, biding its time for the next fishing opportunity. This run across to Panama had, fish-wise, been as dead as a doornail to date.

Since the gas lines and solenoids had been fixed in Antigua, the flames on our stovetop had become cooler. Now my fancy curries and paellas were bordering on stews, which was frustrating. It seemed although the solenoids had been replaced, for some reason, the gas flow through them may have been reduced. Who knows! You know when you get some gas bottles filled with good gas, and another gas bottle of gas will not burn as well because the mix is lousy? No, it was not that; it was the same gas bottle. Worse than that…the fridge/freezer had completely shut down! Effectively, we now had a fridge getting hotter and a stove getting colder! This was just a dent in my pride, as I sought to cook authentic, tasty dishes for the crew in alliance with each country we visited.

We had just stocked the boat in Antigua with quality meats, salamis, cheeses, and veggies. So guess what we had to eat up in a hurry? Food that was supposed to get us through to at least the Galápagos Islands! The UHT milk was now hot, so was the beer and water. We had a big can of butter aboard from Gibraltar, which curiously did not require refrigeration, so we kept an eye on that one. The fact we were moving towards the equator was another challenge. The sticky humidity and heat were getting heavier by the day. The beer cans were swilling about in a bucket of salt water, while the closest to being drunk were wrapped in wet towels. Was Les happy? No!

Bruce drank rum and water; he was okay. Glenys and I are rum and Coke girls. Coke cans joined the dance with the beer cans in a bucket of water in the fore toilet/shower; this seemed to be the coldest place for them to be. The seawater was about 28 degrees outside if you wondered why it was not all over the side.

Copping plenty of sideswipes from an uneven ocean, the hatches were battened down. Sometimes we sneaked them open in our forward head just so we could breathe. Woe betide anyone who left a hatch open in a sea. We did not want a repeat performance of everything being drenched inside the boat! Les and I had a small hatch open in the head. This formed part of the new beer cooling factory. I figured since it was effectively a wet room, the occasional wave on the lee side of the boat would not do any damage. Still, it would make the beer more drinkable. The air flowed from in there through to the saloon and galley areas by opening the door up. I dreaded anything going mouldy as we were heading to the Pacific. Green mould is so difficult to budge on a damp yacht. Thankfully, *Marmax* had been well equipped with Heller fans. Occasionally, Bruce spoiled us by cranking up the generator to run the air conditioning. It was such a relief, especially when we were in the depths of teaching Janelle at *Marmax* College.

By now, we had the challenge of fixing or, perhaps worse, replacing our galley fridge; this threw a spanner in the works. We needed two to three days to sort the fridge problem and restock *Marmax* with fresh food again. Panama was not envisaged to be one of the most hospitable of ports. Still, we were attempting to get into Shelter Bay Marina, just inside the entrance to the harbour. Hopefully, there was an English-speaking refrigeration guy around. We were not really sure what to expect from this destination in the coming few days.

Meanwhile, still at sea, the wind had dropped entirely off, and we had to motor just to keep moving. This was also to avoid that dreadful clanging of sails and rigging when you have a yacht flapping about in a lumpy ocean.

Food, drinks, and plates were flying around the inside of the boat; *Marmax* had worked up a maddening rock-and-roll, side-to-side rhythm. Boat stuff was always picked up, mopped up, and cleaned up as it leapt out of exploding fridge and cupboard doors or off the top of benches. A big tin of chicken escaped from that damn fridge. Bruce's tea and Janelle's lemonade accidentally spilt across the galley floor. To add to the mix, salt water had trickled through the main hatch and a container of sloppy yoghurt had flipped on its side and burst. Poor Glenys was often doing four things at once. If she took her eye off something while doing something else, she often lost track of what she was up to in the first place...like securing a drawer or door. We all stuffed things up by not locking food safely into the fridge or not double-checking a drawer latch containing dozens of knives, forks, and spoons. One often wondered why it was she who was always on her hands and knees cleaning up. At this point of the voyage, we were down to only three porcelain cups left with handles. Glenys sported red knees.

Handwritten messages were placed inside bottles, kissed by Janelle for good luck, and ceremoniously lobbed into the water. The crew was getting serious reading completed on some lengthy novels. We slept when we could. Lifted weights, played music, nursed our huge bruises, which were accumulating, and drew pictures on vegetables for entertainment. Les had bark missing off him from all his knocks. At one stage, he even had to resort to playing beads with Janelle to keep himself occupied until 5 o'clock sundowner time. Nick, my son, had downloaded a hundred or so movies for us in Barcelona to a hard drive. This hard drive turned out to be one of the most precious pieces of equipment aboard *Marmax*.

While travelling this close to the foreboding-looking Colombian coastline, it was time for a pirate briefing aboard. We did not have a complete game plan organised for this sort of danger. It's one of those subjects you don't want to talk about but have to. The anxiety of your worse fears coming true just because you thought about it was dumb. F.E.A.R.: false expectations appearing real. What else do they say? "Plan for the worst, hope for the best," so that's what we did. We were stopping for no one!

We had learnt a few strategies from others before us, so we felt comfortable about it all. One simply did have to prepare for the worst, whether we liked it or not. It was definitely hair-raising and spooky to be gliding along in full moonlight on a dead flat sea, knowing a notorious jungle coastline was coming up. It was so quiet when on watch; you were conscious that a pirate boat could actually sneak up and board you from behind quite quickly. We would have felt a lot more comfortable being approached by an unidentified vessel in 25–30 knots and a four-metre swell!

More birds, fish, insects, and oppressive tropical heat started arriving as we got closer to Panama. Time for a swim. Over the side, four of us went, sheer bliss! Yes...that water was still a fantastic luminous blue; you could see forever underwater. Precisely twenty hours later – yes, we counted them – Les caught a beautiful yellowfin tuna. As he was gill and gutting it and biffing the first lot of guts over the stern, four enormous three- to four-metre-long tiger sharks came zooming up behind us. Oh my god! It was terrifying to think we had been swimming not so long before. We gaped at each other, all wide-eyed and our hands on our hearts. Phew! We could have been four big juicy lures trailing out the wake of the boat; Bruce would have had to sail all the way home alone!

Les had to abandon fishing after that, as it was pointless. These guys were in a total feeding frenzy by now, and the cost of lost lures and tackle had mounted up quite considerably! Speckled dolphins arrived shortly after the menacing sharks disappeared. We spent the rest of the day motoring along on a glassy surface and watched enormous schools of tuna charging about, chasing flying fish; there were acres of them! We had a bit of fun with them.

At 180 nautical miles from Panama, we started getting thunder and lightning. A gentle drizzle of rain wafted over us as we got closer to the Venezuelan Coast. This was only our second lot of rain since leaving Spain three months earlier. Glenys was liaising with our Panama agent, Erick, to organise everything for the Panama Canal transit. You cannot just roll on up to get through this mighty waterway. Everything should have been booked months ahead, except for a marina berth, but not so in *Marmax*'s case; we had not expected to have needed one.

For young Janelle, the *Marmax* College curriculum was under full steam. As her English teacher, I was happy to report we had already completed 90% of her year of work. It was now my job to ensure she remembered it all and kept her antennas up and lively as we went over all the lessons again.

We could not always purchase compulsory courtesy flags before arriving in each country, so we often made them ourselves. Glenys made the one for Panama, one of red, white, and blue squares with two stars on it. She went to raise it as we entered Panamanian waters but then realised she had sewn the string on the wrong side so it would effectively be upside down and back to front. Oops. She quickly undid it and sewed it up correctly just as the customs boat went past. Another hold-your-breath moment behind us!

Early morning dawned, and surrounded by ominous skies, sheet lightning and marching rainstorms, we sighted land and arrived at the entrance to Panama. The first actual city we had seen in three months. Forty-plus cargo ships were lined up and waiting to get through the canal, all shapes and sizes. Motoring through the massive breakwater protecting the harbour zone, we slid through a narrow entrance into a little piece of, totally unexpected, paradise! What a lovely marina facility! The marina guys were friendly, with welcoming grins on their faces. We were so delighted to find luxury facilities ashore; you have no idea how refreshing it feels to have a hot shower after being at sea for days! A sparkling swimming pool, laundry, mini-mart, chandlery, restaurant, and bar were sitting right at the end of our marina arm. This was unbelievable! We had a preconceived notion that Panama would be full of rubbish, oil-slicked waters, and mosquitoes. What we had here were clear waters, full of fish life and squeaky-clean marinas; we even had an air-conditioned library and TV room!

Most of the boats berthed there were heavily laden with ocean-crossing gear on board, with many families travelling from England, France, and South Africa. The mix of accents was fun trying to figure them out. We thoroughly enjoyed meeting so many interesting people and swapping books, clothes, recipes, and experiences. The beer and rums were only US$1.50 to $2 at happy hour. Well again, I guess you know where we were hanging out after 5 pm. We were so thankful our fridge was an easy repair for the local fix-it man on the marina. We were also relieved to have this vital appliance going again.

The lack of English-speaking officials and red tape made life a little complicated here, as were the many power cuts we experienced. Still, our time there was most memorable and well beyond what we had expected. The marina was located on the edge of a jungle, seriously, in the middle of nowhere. When you were in the pool, hooded vultures sat in the palm trees watching over everything with their mean, beady eyes. Early in the morning, you could hear the howler monkeys sounding like a kennel full of barking dogs. The rainforest wrapped around the marina complex; it was great to listen to birds again! The Caribbean had sadly lacked this due to there being no trees after the hurricanes.

The third day in, we clambered aboard the free Shelter Bay Marina bus to be taken into the city of Colón. Safety in numbers, they chanted! It was a twenty-seater coaster type vehicle. We lurched and swayed our way through a jungle road bordered by lush growth and sugar cane. Much to our astonishment, we discovered we were being protected by a fenced compound guarded by an army security outpost! No wonder it felt good in there.

An old, rubble road led us through a derelict, abandoned accommodation precinct. Bizarrely, our bus then popped out on a new road development, which was being put in place for a large suspension bridge, already proudly stretching across the canal. It was massive, impressive, and so seemingly out of place in this lower socioeconomic environment. When we got on that bus, we had no idea what to expect. After fifteen minutes of travelling, out of the blue, we found ourselves suddenly driving across both of the Panama locks. While gazing up at the colossal cargo ships above us as we went by, the world right then seemed full of surprises.

Have you ever been to Colón? It is the type of place you would be curious to visit once in your life but a relief to leave. Half of our fellow yachties had gotten out of the bus at the local supermarket. But always up for an adventure, we stayed aboard to go to the Free Zone. Sounded dangerous, um…yes, it was. Until we had arrived at the guarded entrance with armed security buzzing everywhere, we didn't know the Free Zone was the duty-free shopping zone. Duh! We had to show passports and go through an intense security procedure to get into the shopping area. It was full of dilapidated commercial buildings with enormous security grills, guards, and guns everywhere. When you went into a shop, you had to leave your small backpacks or bags with a guard at the door, or you could not go in. You were not allowed to try clothes on, just buy them. Passports were shown for every Visa transaction. We were electronically wanded outside the bank, then the bank was unlocked with a key so we could get in. Another guard was on the other side to check us out. The place was very cheap, but then again, you had to buy everything in bulk. Now we had a full appreciation of how inflated our prices were back home.

All the big brand names were on show, loads of wholesale shops for merchants, lots of gorgeous traditional food stalls cooked from open fires and under various shelters in the streets. There was very little English spoken or written. It drove us round the bend, having little bland-faced shop assistants fluttering about us. They stood guard and positioned themselves only a foot or so from us the minute we walked into a shop. Did they not want our money? The streets were filthy, but now and then, you could score a smile from someone in the street. It was an enclosed city, worth seeing once in your life and a super exciting, eye-opening shopping adventure for us all.

Leaving the Free Zone was disturbing, dangerous, and at the same time, hilarious. We had to catch a taxi back to the supermarket to link the bus returning us to the Shelter Bay Marina. There was no way you would walk outside the guarded compound. We were advised that a taxi would sweep in alongside us if we just adopted a useless "I'm a tourist" look; it had proved impossible to flag one down. So this tiny, beaten-up Fiat stopped to pick us up. The 120+ kg taxi driver was an albino native, a heck of a nice guy. There were six of us who needed a ride. We had somehow collected a South African yachtie who wanted to stick to our group; his name was Johnny. Six people could not fit in a Fiat, or could they?

"Jump in," he gestured. English was not an option here.

What?? Six of us plus our bags? Our bags got jammed in the boot. Les sat in the front seat, while Johnny, Bruce, Janelle, Glenys, and I crammed into the back seat, all in hysterics! We had to go through the military-style security checkout. Stone-faced guards unrelentingly stared in our windows to check our faces. They had no problem with looking straight at Glenys's backside, which was jammed into the rear window; they must have been wondering where her eyes were. That was a pass. The guards opened the boot and checked it out. All good to go. Phew!

We found the great supermarket we had all been advised by the marina to shop in. It was cheap, cheerful, and full of friendly staff and every brand imaginable. It was clean and had the first decent butchers we had found since at sea. What a land of contrasts! We filled half the bus up with our groceries. The rest of the yachties arrived soaked to the bone from a torrential downpour in action.

Merrily, we took off to return to the marina, a forty-five-minute ride through what was now mud most of the way. The driver told us to buy a lotto ticket. We had scored a safe bus passage through both locks on the Panama Canal again, without having to wait an hour plus as the other buses had. Yay!

Our canal transit was booked for the next day; *Marmax* was ready to roll. Her hull had been measured, and we had paid the canal passage invoice. Extra lines had also arrived and were piled up at the marina. Good Lord, they were long, heavy ropes compared to what we were used to! Large fenders had to be rented to enable us to protect the hull of *Marmax* for the passage.

It really felt like we were heading home now. We were leaving the Mediterranean, the Atlantic, and the Caribbean in our wake; what a fantastic adventure it had been so far! The next ocean we were to sail through would be our home waters, the Pacific.

One major drama for us was that we had great difficulty getting official authorisation to get into Galápagos, our next destination. Glenys worked frantically to get paperwork off to the yacht service agent, who would hopefully smooth the way for us. We had just found out we should have booked at least two months ahead, not two weeks. The agent did not seem at all confident we would be allowed in at such short notice. Despite his lack of faith, Glenys emailed through our application anyway. We would cross our fingers that our lucky streak would continue. As a last resort, we considered visiting the Ecuadorian embassy in Panama City once we got through the canal locks. Like the Panama Canal, you could not just turn up in the animal kingdom of the Galápagos!

Foreboding weather off the Venezuelan Coast

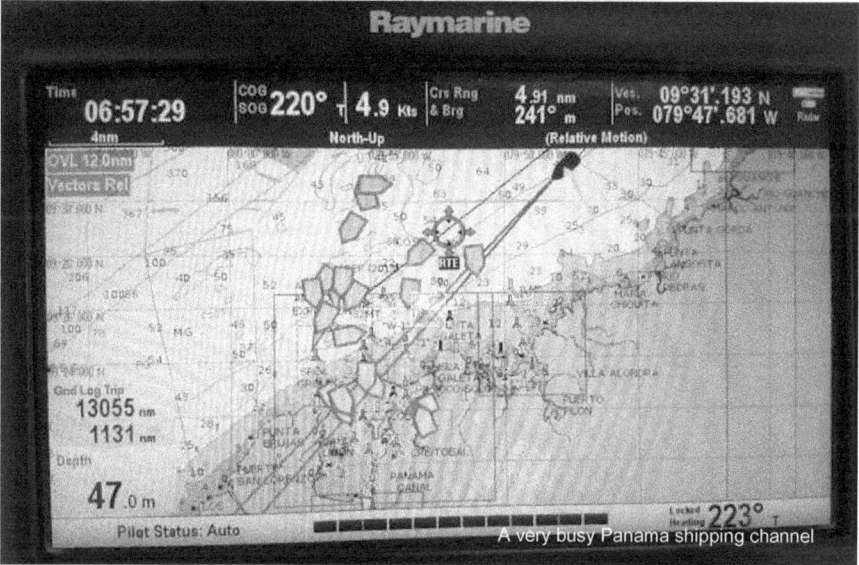

Raymarine

| Time 06:57:29 | COG 220° T SOG 4.9 Kts | Crs Rng 4.91 nm & Brg 241° m | Ves. 09°31'.193 N Pos. 079°47'.681 W |

4nm   North-Up   (Relative Motion)

OVL 12.0nm
Vectors Rel

Gnd Log Trip
13055 nm
1131 nm

Depth
47.0 m

Pilot Status: Auto

Locked Heading 223° T

A very busy Panama shipping channel

Entering Shelter Bay Marina, Panama

Glenys sewing the Panama flag

11. Panama, Las Perlas: We Get Clearance to Enter Galápagos!

To conquer and experience the mighty Panama Canal was an absolute thrill for us all.

Besides the captain, we had to supply four experienced line handlers to work the four heavy forty-six-metre lock lines. These were necessary to reach the high sides of the canal from the boat. The powers-that-be declared Janelle was too young, despite her vast boating experience. So young José, our new line handler, arrived on *Marmax* at Shelter Bay Marina at 1 pm, the day of the start of our Panama Canal journey. José had assisted small vessels through the locks over three hundred times.

"Hey, José, do many people swim in the ocean off Panama?" we asked.

"Ohhhhhh no! Nobody does that. Stupid. Get eaten by a shark, hey," he replied with his forehead puckered and a look of horror.

"Yes." We nodded knowingly; we would not be that stupid, would we??

Relaxing on the top deck, we enjoyed a spectacular rain and electrical storm show whilst waiting. The afternoon had become dark and overcast. Time was ticking on; we were surprised to learn we would not begin transit into the canal until at least 5.15 pm. It did seem a little late in the day!

Our pilot, Ivan, arrived on a tug alongside. All boats had to have pilots aboard, and Ivan was the leading man for our flotilla of three yachts rafted together to transit through. We were allocated the central boat position, with a 49-foot French yacht *Sandetie* lashed to port. The tiny 27-foot USA yacht *Tarka* was secured on our starboard side. *Sandetie* had travelled worldwide for twenty years and had been to New Zealand and Australia twice; it was their third time through the canal. We were in good company here! They were on their way to French Polynesia.

Tarka had already circumnavigated the world with a previous owner. The current owner, a lovely young man, had bought her in St Maarten in the Caribbean. Being new to yachting, he had just spent two years sailing her from St Maarten to Panama. Funnily enough, *Marmax* had just sailed that same passage in only ten days; he could not believe we had moved so fast! The speed at which we were travelling was the source of amusement for many along the way. But here we were, right on time! The hurricane season was beginning; our insurance policy required us to be in Panama by the beginning of June.

Once all three pilots had been safely dropped off on their allocated yachts, we proceeded in single file to an area just before the locks to all raft up tightly. It was such a different world these guys alongside us lived in, real sea nomads! It was then we all got to meet each other, shake hands, and learn about each other's adventures.

Panama Canal is considered one of the greatest engineering wonders of the world. Columbus started cruising Panamanian waters in 1502. Years later, Panama became a transhipment point for gold and silver from Peru. Due to the tricky terrain, it made it a long, arduous trip getting goods from the Atlantic to the Pacific side of Panama. In 1534 Charles I of Spain ordered the first survey of a possible channel going from the Atlantic to the Pacific side of Panama. The French were first to start digging in 1882; however, the project was abandoned, as it was too costly and dangerous, and many workers died from tropical diseases.

In 1902 the Americans took over with the proviso that they could have permanent control of the canal. For two years, thousands of workers cleared bush and drained swamps. Many died from malaria and yellow fever from mosquitoes, and bubonic plague from the rats. It took ten years and seventy-five thousand people to build the largest locks in the world and complete the canal.

Before its grand opening, any ship going from New York to San Francisco would travel 11,000 nautical miles. The canal shortens this by over 4300 nautical miles and avoids the treacherous Cape Horn. The first passage through the canal was made in August 1914. In 1979, the canal zone and its administration were given back to the Panamanian people. Ownership of the canal was transferred to Panama in 1999. An expansion of parallel locks to allow oversized ships to pass through and cope with the rising demand for transiting boats started in 2007. This took nine years to complete. This has been a massive money earner for the country, which is now thriving.

High on buzzing adrenaline, we entered the first Gatun locks. Four canal attendants, high up on the wall, threw monkey fists to our yachts. Monkey fists are a type of knot, so named because they look like the clenched fist of a monkey. They are tied to the end of a light rope, or heaving line, to serve as a weight. The monkey fist is thrown to the boat, enabling the crew to attach heavy lines to it, so they could be pulled up to the top of the canal wall. Usually, fore and stern lines are secured to each outside yacht. We ended up having the ropes attached to us, as tiny *Tarka* would have had her cleats ripped out of her hull if she was to take the weight of all three yachts in the locks. The central yacht position is apparently the safest. When the lock controllers raft three or four boats together, the two outside vessels usually hold the entire assembly together. If their lines fail or a line is mishandled, this usually causes hulls to crash into the walls. Smaller yachts like *Tarka* can hit turbulence and roll far enough for their spreaders, on their masts, to hit the wall. Scary stuff!

It was absolutely fascinating to see history etched into the sides of those concrete walls. The old fixture points where ships, from decades past, had attached their lines and the infrastructure of the old locks were still evident, as were the gouges of ship gunnels that had been scraping past.

Once we were secured to the walls by the linesmen, the massive concrete gates were shut, then the locks were filled with water. The gates were later reopened, and we moved into the next lock, then the next. The three chambers of the Gatun locks had raised us eighty-five feet up to the level of the freshwater lake of Gatun. Highly organised, professional, and totally impressive. However, for us, it was not only exhilarating but very stressful!

The little linesmen (they looked small, as they were so high up on the walls!) wore heavy uniforms and big boots with necessary bright orange flotation boards strapped to their chests, both front and back. They were often trotting along burdened with all that heavy gear, heaving ropes, operating radios, and stepping over bollards and steel bits. I asked our pilot how often they fell into the water.

"Often" was the short answer. An answer I had expected.

Even more fascinating was watching those big "mules" at work. The shining, state-of-the-art locomotives hauled and controlled the lines connected to the large cargo and cruise ships.

My second question to the pilot was, "How come these mules don't fall into the water when they are only sitting on tracks?"

Answer: "They do fall in, but not often."

Wow, that would be costly! When the mules pulled ships from lock to lock, they tried never to take the entire load unless pulling forward. Tugs, four-million-dollar power machines, followed the vessels to help the mules control the movement. The mules gently pull-stopped-pull-stopped so as not to lose their balance on the track nor allow a ship to move too quickly. They were wonderfully efficient, the drivers were all grinning, and a lot of waving was going on. It was incredibly nerve-racking to feel so small under the bow of a giant container ship sharing the same lock as *Marmax*, twenty-five metres above sea level. You felt like a mouse about to get eaten by a cat!

Tragedy stared us straight in the face. As the water was rapidly swirling into the first lock, a young deckhand on the French yacht alongside us, unthinkingly, started pulling in the slack of his forward line. This line had the full load of the three boats that were attached. All three yachts immediately lurched sideways towards the starboard concrete wall. Thank goodness, with all the yelling going on, the young man suddenly realised his mistake. He quickly recovered the situation, not without a wicked hand burn. He could have easily lost his hand or fingers. If we had hit the wall, *Tarka* would have been toast and would have acted as a fender for our 21-tonne *Marmax*!

The Panama Canal transfers ships from one ocean to another via a series of locks, for those unaware. The river Chagres was initially dammed to create a lake that holds the water necessary to operate the locks. Propelled by gravity via a system of culverts in the central and sidewalls, the water runs under the lock chambers. Each lateral culvert has five holes measuring 4.5 inches in diameter. The floor of each lock has one hundred holes. The pressure of the water is immense. It took around fifteen minutes for each lock to fill. It was all action! An incredible fifty-two million gallons of freshwater is needed to lock a ship from one ocean to the other after doing its work.

Despite the heat, humidity and rain, the sunset was magnificent. Our raft of three emerged from the third lock safely, with a great deal of back-slapping and handshaking with our new friends. After untying our raft, in single file, we slid silently through the fresh waters of Lake Gatun, past the exit of the mighty Miraflores locks. We secured ourselves for the night on a simple, circular, mooring buoy. It was about three metres in diameter with soft plastic sides so as not to damage any vessels. *Sandetie*, *Tarka*, and *Marmax* tied up to the central mooring hook on one; two other yachts from another raft ahead of us tied up to another.

Unfortunately, we were all too hot and exhausted to have drinks with the other boats that night, but I bet those buoys could tell a few good party stories! After a fitful sleep, despite the oppressive humidity, I awoke at 5 am, as I wanted to listen to those howler monkeys roaring one last time. I did not doubt that Tarzan's famous cry was initially inspired by the howler monkeys. They seemed to really get stirred up just before the sun rose. I'd hate to get caught in the forest at this time with them; their howling was the stuff of nightmares!

The young crew aboard *Tarka* were stirring by 6 am. Lord knows where all six people slept that night on the tiny thing! I heard a splash. One of the French line handlers had gone for a swim.

Les yelled out to him, "Mate, did you know crocodiles are swimming about in the lake?"

He jumped out with a smile on his face. We were not sure if he knew what we were talking about. Another deckhand accidentally bumped his head on a swinging net basket holding a bunch of bananas on the stern…yes, I said bananas. They went flying into the water. Yikes! Ever seen a Frenchman swimming after a bunch of floating bananas in a crocodile-infested lake? My eyes were glued to him until he got out of the water. It was a quick swim!

Our pilots had left the boats the previous evening, and a new pilot crew were due on board to guide us through the next part of the transit. At 8 am, the pilot boat came roaring alongside with four men aboard, two pilots plus two trainee pilots, but soon we learned none for *Marmax*! We were, therefore, left alone on the mooring for forty-five minutes until our pilot and his accompanying trainee turned up. We were not happy, especially considering the cost at which Bruce had forked out to his agent to ensure all went well. You simply did have to go with the flow and trust the pilot was actually going to turn up. Very few of them spoke English. Luckily, we had a strong engine, as we had to catch up with the others who had a head start on us.

We steamed thirty-one miles up the tropical waterways of Lake Gatun. Our wonderful, pristine, sapphire blue Mediterranean, Atlantic, and Caribbean Seas were behind us. We were now in a murky, freshwater environment, with old treetops still poking up from the water from the days when the lake had been flooded years before. There were lush Amazonian-like rainforests growing all the way to the water's edge, abundant birdlife like we had not ever seen on our travels, and crocodiles cruising past. And oh my goodness, it was hot!!! The air was so stifling, we could barely breathe. Mercifully, the movement of the boat, making her own breeze, was a relief! A magical, tropical environment, full of islands, creeks, and inlets, volcanic mountains, and 120 or so port and starboard navigational buoys showing the way.

Stopping off for lunch on another one of those innovative mooring set-ups, we enjoyed quite a show of fast-moving, massive ships passing us. They all had tugs bridled onto their sterns as they followed them through to the next set of locks. We were sitting there to get out of their way so they could reach their scheduled transits; we were also waiting for the other yachts to catch up. One of these fancy tugs gave us an incredible water squirting display; there was plenty to watch.

Off we continued, under another spectacular suspension bridge full of traffic seemingly linking one rainforest to another. Where was everyone going?

It was time to raft up again. This time, with a Turkish 48-footer, *Ginger*, with a father and two sons aboard. They were off for a long-awaited, eight-month dream cruise to French Polynesia. They had many foreign volunteer crew members aboard, helping their yacht through the locks to make up the four-line-handler rule. Later on, the handlers all jumped onto a water taxi and then caught the bus to return to Shelter Bay. There seemed to be many people willing to help out on these transits, and obviously, it was a great way to score a position as the crew to get to the Pacific. We had a couple of friendly guys ask if they could join us, but we shook our heads. "Family only." It seemed a good reason. So many of these guys smoked cigarettes; you simply could not have one on board if you were non-smokers. Sadly, smokers were discriminated against when trying to get positions. As usual, Glenys, the great Samaritan, tried placing a couple of guys while we nattered away to other yacht owners. The fact they smoked was the only reason these two could not get aboard these beautiful big adventurous yachts.

The following three locks down were easy travelling now as we pretty much knew the system. We dropped thirty-one feet in the first Pedro Miguel lock then steamed one mile to enter a double lock that lowered us fifty-four feet down to the Pacific. The feeling was euphoric! Looking over the top of the gates of the final lock, there laid the Pacific Ocean. Out with the blue and in with the green. The water looked like pea soup by comparison to our dream run across the last three seas. We all heaved a big sigh. The Panama Canal had been conquered, a lifelong dream for many was complete. We wondered what the home stretch had in store for us!

On a perfect sunset, and with an escort of huge brown pelicans and colourful parrots, we slid under the Bridge of Americas. It was almost a replica of the original Auckland Harbour Bridge in New Zealand. After radioing ahead, we picked up a mooring in the harbour belonging to the Balboa Yacht Club. What a brilliant, exhilarating, fascinating and challenging day! It was still hot as hell, but by now, at least, we had a cooling sea breeze coming in from the Gulf of Panama.

We spent the night on the fringe of the busy shipping channel. The Panama Canal operates 24/7 for the big ships, so we had shadows of enormous freighters all night and the noise of their servant tugs charging back and forward. We were so tired, we barely noticed.

Our main reason for stopping there was to clear customs and immigration in Panama, but while here, we thought we may as well go for a reconnaissance of the town. A complimentary water taxi man was operating on the moorings. It was a quick whistle, a shake of a hooter, or a call on the VHF, and he would be at our side. They were so efficient! One thing that did freak us out was when these big, bulky fibreglass taxi boats came pounding up alongside our shiny yacht hull. We all took a big gulp and held our breath as they came near. As a Kiwi, I realised that we tend to treat our yachts like precious pieces of jewellery by comparison. Almost every Panamanian vessel was simply a workhorse. Those things bang, smash, and clonk into anything and everything, but boy, they get some work done. This was not the sort of port you would throw down your tender and outboard to get ashore. Besides, ours was still safely deflated down in the lazarette with the outboard to avoid any temptation from local undesirables who might fancy them.

Have you have ever seen the television show "The Amazing Race"? If so, you will realise what it is like to travel as Team *Marmax* when ashore. We had to get two taxis into town because there were five of us, and well…five to a taxi was, simply, the law. Including the driver, equalled six. After we had hailed one down, that driver, in turn, hailed another for us. We were taken on a wildly exciting route through outrageously coloured shacks piled on top of one another among piles of rubbish. We got road blocked by a singing and marching band of people in traditional dress and then diverted by the military and police for some sort of cyclist race. Miles of colourful national flags stood to attention along the impressive waterfront. Modern buildings, deserted embassies, and sports stadiums were strewn among the derelict, once-grand mansions. We were told they were relics of the long-gone American army; it was all a little sad to comprehend.

We were very keen to check out "Old Town," which had just celebrated its five hundredth anniversary. To get to such a destination, one had to speak into a translator app on your phone and show the Spanish- non-English-speaking taxi driver where we wanted to go.

For those who do not know how they work, he had to speak back into it in Spanish. Teaching our driver how to talk into the phone over the din from his music blaring was amusing. There was much hand gesturing, smiling, nodding, head shaking, then clapping. He understood! Phew!

It was well worth the ride into town; we had a ball! The Panamanians were going hell for leather restoring their charming old buildings, which had survived time, but only just. Though not quite as elaborate as in Europe, the churches were grand. Les got the opportunity to have a chat with the patron saint of fishermen, Virgen del Carmen. He thanked her for his fishing abilities and asked permission to keep having fun. I was pleased to note he also asked her to bless me amongst all his fervent whispering. She was pretty extraordinary. Her statue plaster eyes seemed to follow him around the church even as Les sidled out the door; I swear she winked at him on the way out!

There were some rather astonishing bronze statues, many of them depicting babies, children, women, and war. As you would expect, every one of them had heroic stories in Spanish, which we took the time to decipher on our translator app. The usual colourful markets, but this time, many were authentic Panamanian artisan work. The art galleries were uniquely superb. The buildings? Well, I was happy as a pig in mud taking photos of such beauty. Let's say they had not yet succumbed to the blandness of tourism nor lost their direction while embracing their history and artistic talents.

Of course, Panama is the home of the Panama hat; who would imagine you could also get pink and green Panamas? There were no prices in many of the shops. Still, everyone was friendly despite not speaking our lingo; they were great barterers and gave us reasonable prices. Les and I indulged in a rather romantic lunch in a great restaurant called the Cosmopolitan. It also had the coolest air conditioning, excellent wi-fi, and icy beer! Partaking in a meal of delightfully super crunchy pork crackling and a sensational clam broth, we finished off with chicken baguettes. We delighted our friendly, attentive waiter with our compliments. It was not only the best restaurant meal we had had in months but the very best service ever experienced. Simple food but perfection. When on extended yacht cruises, you do miss tasting food with a good crunch! At every corner, the Panamanian cuisine looked delicious; this was something we had definitely not expected.

Unfortunately, we could not dither, as we had made a plan to meet back at the Balboa Yacht Club by 5 pm. It was no surprise to us to read that Panama tied with Paraguay as the planet's number one happiest destination that year in the Gallup Global Emotions Report.

We enjoyed a few drinks at the yacht club: a thatched roof with a beaten-up bar, a book swap, two broken toilets, and no staff who understood English. Because customs and immigration located alongside the yacht club was closed on Monday, we stayed a second night in the harbour. The following day, we came ashore to clear the country. In we went to the Balboa Yacht Club. Tempers got short with the official not knowing a stitch of English. Finally, he told us we had to go to another office at the Flamenco Marina, miles away, via that handy little translator app. We had it on good authority Balboa could do the task, but today, it seemed not. This conversation took over an hour to extract from the guy.

In the meantime, an overly friendly taxi driver was badgering us to jump in his taxi nearby. How many times can you tell someone the word no! As it turned out, all five of us (forget the law now) ended up in Louis's taxi, a beaten-up affair with deafening music. At least twenty cardboard tree fresheners hung off his rear vision mirror. His safety belt was connected to the car with clothes pegs. With a cocky wink and a devilish smile from our driver, we took off like a bat out of hell to the Flamenco Marina, fifteen minutes away.

This was a much larger customs and immigration department, but still no English was spoken. The trusty translator app came out – it was getting quite ridiculous!

Here we were, the Marmax crew all lined up like naughty school kids with our bags on the floor, getting scrutinized by two very surly uniformed women. We may as well have arrived from the moon! Fortunately, a cute-looking Costa Rican guy lit up our morning by saying hello to us in English. Hooray! A translator! He was clearing his boss's boat, an 86-foot luxury yacht. Bruce got marched off to another room for interrogation while we endured the third inspection of our papers and passports. We waited, and waited, and waited along the wall of that customs office. Louis, the taxi driver, would not leave us. He insisted on waiting outside. After a considerable length of time waiting in customs, Glenys went downstairs to attempt to pay off Louis.

"Please go, Louis! We don't need you! Understand?"

Again, it was translator app time. He ended up following Glenys up to customs; clearly, he was well known to these official ladies. Glenys finally got rid of him. Bruce came back relieved to have gotten clearance. They had Marmax's photo on their computer. Still, they were puzzled that they could not pick up the AIS (automatic identification system) position, as Bruce had turned it off in port. All explainable. Clearance papers were ready to be handed across.

"Where's the taxi driver?" they demanded.

"Gone," we gestured with our hands.

"You sent him away????" Sparks of anger flashed out of the ladies' eyes.

"Yes!" we nodded vigorously.

Oh dear, we had some distraught customs officials on our hands – the look of horror in their eyes! Priceless! It seemed we had made a mistake in sending Louis away. Fortunately, Glenys had the forethought to get his mobile number. Customs rang it, berated him for leaving us, and the next thing, he turned up in the office with a beaming smile on his face. The clearance papers were given to Louis, not Captain Bruce. We were escorted by an official-looking fellow down to Louis's waiting taxi at the door. All five of us were hustled in with an armed guard standing over us. We were driven directly to the Balboa Yacht Club, despite Glenys pleading with Louis to let us stop along the way so we could buy potatoes and bread. Louis handed our papers to the other stone-faced customs official in his little office next to the yacht club with his chest now puffed up. It was made very clear to us that we had to leave Panama immediately.

We refuelled at a very precarious-looking diesel dock. Broken steel spikes protruded from the concrete pontoon, threatening to gouge out the sides of our hull. We had every fender over the side, protecting us from the ferocious-looking platform; everything became covered in grease and oil. The fuel came out of the bowser at full pressure, creating a blowback of diesel straight up into Bruce's face. Let's get out of here!

Motoring out of the harbour, we were again escorted by hundreds of massive brown pelicans zooming across the water and soaring up in V formations. With the sails eagerly hoisted, we headed off to Las Perlas Islands, some 30 nautical miles away. Plus, we had just received notification that we were now cleared to enter Galápagos. Were we excited? Oh, yes, we were!

Las Perlas Islands are nestled like a pearl in the Gulf of Panama. We arrived in the dark and anchored off the stunning Don Bernardo beach, located on Isla Pedro González, and what a beauty she was!

A long white sandy beach, something that everyone dreams of in a remote tropical island. Coconut and mango trees dripped with fruit to the sandy edge.

Our tender was still locked up in the lazarette on our stern, so we blew up the kayak, and Janelle ferried Glenys, Bruce, and I ashore. You see, technically, we had already cleared this country, so we risked a very brief visit. We were not too comfortable about the shark and crocodile situation here, so we played with caution. After some beachcombing and collecting mangoes and coconuts, the heat got the better of us. Bruce, Janelle, and I nervously swam back to *Marmax*, a fair way out off the beach. Hang it; it seemed safe. We had been observing several native fishermen in their little boats, pulling up ashore to feast on mangoes for lunch; they ate the mangoes skin and all. A lone boatman had waded around in the water tendering his boat while waiting for his mates to return. We swam boldly out to *Marmax*, escorted by soaring pelicans. Swooping down, they looked at us curiously with their heads tilted sideways, as if they were short-sighted and needed a closer look. We made it back safely and alive, apart from Janelle copping a painful whipping from an unidentified jellyfish. I'd like to think the pelicans were guarding us.

In the afternoon, we lifted anchor to go circumnavigate Isla Señora, a pelican breeding sanctuary. It was a fairly rough sea, as the wind was blowing hard. Glenys kayaked ashore alone to check out a bay where baby pelicans were learning to fly. The rest of us stood on standby aboard *Marmax*, keeping a close eye on Glenys out in wilderness we were not familiar with, in a pitching sea. We knew the pelicans would go into flight mode once they met Glenys, which they did… in their hundreds! It was a spectacular sight!

On our way back, we passed a native village on Isla Pedro González. The colourful coastline was alive with tiny houses and streets dotted with red, blue, and green satellite dishes. A few white memorial crosses shone in the sun. Dozens of traditional fishing boats lay on their sides on a palm-fringed sandy beach. Sadly, the village sat within a circle of household rubbish. It had been disposed of over their front fences, tumbling down onto the rocks bordering the sea below. This was our first disappointing experience of coming face to face with human pollution on our entire trip. I mentioned earlier the land of contrasts. One mile away, we discovered a brand-new, forty-five-berth marina development with four luxury launches moored, topped off by a salubrious boat club presiding over it. This was phase one of the new Pearl Island Marina development. Unbelievably, the second phase was planned to have a capacity of three hundred boats. It seemed bewildering to us, as it seemed so out of place here in such a remote, primitive location.

A super show of rocketing dolphins zooming vertically up in the air entertained us on the way to our next anchorage, Isla San José. This island was only three hours away and a completely different environment from what we had just left. We spent the night hanging between two beaches, both filled with smooth, round, black rocks with no sand at all. It was weird. There was one coconut tree in the anchorage. The whole island was covered in some sort of smothering vine. We had come to this island because we reckoned there would be no rivers for crocodiles to inhabit on this side. After studying the topography of the island, we had come to this decision. The crocodile-infested waters of northern Australia were very familiar to Les and me. We were well aware you could never be too cautious with these prehistoric beasts!

Glenys and Janelle went off for a paddle in the kayak and came flying back like Maori's racing in a waka; a crocodile had crossed their bow! We had been studying the charts looking for a safe anchorage so we could all jump over the side and clean our waterline and hull. Everything below water had to be immaculate to gain authorisation into Galápagos. The topsides had also gotten quite grubby coming through the canal, with many other boats rubbing up against us. Fenders and staining froth had left their marks. Until Panama, we had kept the entire boat spotless. Now here we were in the land of sharks and crocs, faced with having to get into the water. If it was not for Galápagos, it would not have needed cleaning.

By now, it was blowing 17–20 knots. We motored about for another anchorage; the wind was on the nose. Ocean swells had become too violent to enter the water there, so our sails were hoisted, and we took off to check out a tiny, sheltered cove at Isla Del Ray. This was just like a scene from a Hollywood movie set. Very pretty with its three beaches surrounding a horseshoe-shaped bay. Calm waters, lush, tropical, and – this time – pure black, fine sand. Weird!

We donned makeshift stinger suits to prevent getting stung by jellyfish, and four of us leapt over the side, working quickly to clean *Marmax*'s hull. The water was murky here, so we posted Janelle as the shark and crocodile spotter on the top deck. There was no way she was going to jump in among jellyfish. There was nothing to report, but Bruce looked very attractive in his stinger suit, compliments of Glenys's wardrobe. Let's say it was good entertainment!

While lunch was being made aboard, I slipped ashore in the kayak for some photos of this gorgeous little spot. From the yacht, it looked totally pristine. My heart leapt into my mouth when I got ashore! There was a rise in the beach, edged by native hibiscus and mango trees dropping their fruit straight into the hot black sand and cooking them; the smell was mouth-watering! On top of the rise, I was horrified to find piles of household rubbish under the trees. Mainly plastic bottles, flip-flops, and hundreds of croc-type footwear. Wow, it would only take one day for villagers to clean this mess up. Why, oh why, did they allow this irresponsible behaviour? I miserably took my photos, returned to the boat, then we were off, straight into a 20-knot wind. Nice. It was going to be a great sail down the Colombian coast by the look of it.

After we had reefed the sails down, I was sitting, wondering what this 1300 nautical mile adventure would offer us. I looked up and immediately discovered our wind indicator had disappeared off our masthead. It had been there five minutes before; I even had the photos to prove it! Glenys and I loved that instrument. Was it a flying pelican or the mounting winds? Grrr! We could live without it, but it was an annoying loss.

We estimated we would arrive in the Galápagos Islands on our fourth day of sailing, hopefully, Thursday, June 6 or Friday, June 7. Initially, this had not been the joyous ride we had expected. We had dead calm, airless hours of floating about. Within fifteen minutes, it would blow up to 25 knots, so we were constantly changing the configuration of the sails. Some nights, the wind shift was an immediate 180 degrees, disorientating on an ink-black night with no stars! The fluky weather was no surprise. Galápagos Islands sit right on the equator and astride a merging Panama Flow and the cold Humboldt Current from the south. The southeast trades are the prevailing winds, with plenty of rain squalls upsetting the regularity of a good sail. There were no heavy swells, as in the Atlantic or Caribbean, but very choppy seas. Still hot as blazes, the yacht was full of wet-weather gear and clothes drying out. We had been thoroughly spoilt for perfect weather before this trip, so it was payback time, I guess.

Day 6, oh joy! A steady 17–24-knot trade wind had arrived. It was so dark out on the water, still no moon or stars. *Marmax* galloped forward like a giant white horse with phosphorescence coming out her nostrils, flowing down the sides of the boat. There was no horizon, just blackness with the water seamlessly joined to the sky. It was so surreal doing watches at night…

Plenty of action followed from here with the wind and ocean movements, but the boat hatches had to be locked down, making it so hot downstairs. *Marmax* College had doubled the workload for Janelle. Rough seas caused the boat to constantly lurch and bounce about, whilst up on deck, dolphins and booby birds put on impressive performances for us. Full marks were given to Janelle for her newfound ability: concentrating while the weirdest things were going on!

One example: Glenys was upstairs reading the book *Knockdown*. It's all about the tragic Sydney Hobart Race of 1998. She was reading a passage aloud to the boys in the cockpit. The story was about these two rookie paramedic girls performing their first sailor rescue from a chopper in atrocious sea conditions. I had just finished studying a passage from a logbook of the tragic Mount Cook rescue of Mark Inglis and Phil Doole in 1982 for Janelle's English lesson downstairs. Janelle's task was to accompany a helicopter pilot in a sea rescue. She had to write the sequence of make-believe events in a logbook. *Marmax* was on a 45-degree sailing angle thrashing through waves. This was all going on at once, in real-time, to the minute. What happened next?

Bruce yells out, "Chopper overhead!"

What the? We had not seen a helicopter since we left New Zealand! Besides that, we were hundreds of miles from anywhere! How could it be? Anyway, that's what I mean by weird things that happened all the time. We clambered up on deck, waved at the Ecuadorian military chopper. A frigate ship was on standby in the distance; they appeared to be keeping an eye on us. It was time to get back to our lessons.

Janelle's learning was becoming something out of this world. It had been a great idea to take her out of school for a year. The amusing booby birds had just joined us aboard. The dolphins were a great joy and a source of constant excitement to us all. Thousands and thousands of beautiful things!

On our final night before getting to Galápagos, the wind dropped right off. The skies were brilliantly clear, with a quarter moon doing its best to light our path. I came on watch at midnight. We were now motoring on autohelm, so my wheel watch was a doddle. We were constantly scanning the horizons for other boats, as many of them did not appear to have radars or AIS.

I nearly choked, boggle-eyed! There was a starboard light rising right alongside us! Surely not. Everything flashes through your mind in a panic. As it was night, we now had our bow navigation lights on as we were motoring. The starboard light flew away. It was a booby bird, its white chest reflecting bright green, crisis over! I swung my head around quickly; there was another flying portlight on the other side. The navigation lights were swooping everywhere. Hilarious! The bow looked like a Christmas tree; the boobies were loving it! My heart resumed back to a regular beat.

On Thursday morning, the Milky Way sat at the top of the mast as we slid over the equator at 3.39 am.

Marmax was once again spring cleaned in readiness for Galápagos customs. School was out for Janelle for a week. We had waited all our lives to get to the Galápagos, and now we were nearly there. We all had high hopes that it would be the *icing on the cake* for our fantastic sailing adventure across the seas.

Gearing up for the Panama

THE PANAMA CANAL
CANAL DE PANAMÁ
SHIP IDENTIFICATION NUMBER
6 0 1 7 8 2 4
KEEP THIS CARD IN THE WHEEL HOUSE.
YOU MUST REPORT THIS NUMBER ON ALL ARRIVAL
MESSAGES FOR TRANSITS OR PORT CALLS

Crossing the Panama in a bus!

Thursday, June 6, 2019

12. The Galápagos: A Paradise Within a Paradise!

If you have a bucket list plastered to your fridge or wall of your office, you MUST see the Galápagos Islands! Wow! Wow! WOW! You could not wipe the smiles off our faces! It is easy to see why they call them the enchanted islands.

"Land, ho!" yelled Les excitedly. Glenys accidentally lost a precious bucket over the side a moment later, just as we entered the Galápagos territory waters. It ever-so-slowly sunk as we agonisingly watched it disappear into the depths. Why did we not dive over? Because we did not know what was lurking about, and after that shark experience near Panama? No. One must check out the territory first. The family's safety was our number one priority. Silly us. In hindsight, we had never swum in such a fish/animal-friendly environment, but it was better at that time to be safe rather than sorry!

It took us thirty hours of motoring to get our first glimpse of one of the islands that make up the Galápagos. There was simply no wind, but on the bright side, we got to get up close to study the walls of mighty Kicker Rock, also known as Roca León Dormido (Sleeping Lion Rock). The magnificent, standalone rock, some 148-metres high, welcomed the *Marmax* crew. This was our first experience seeing frolicking sea lions, basking turtles, giant frigatebirds, and a selection of animals we had never seen before. The soothing voice of the great David Attenborough played in the back of our minds as we found ourselves wondrously staring open-mouthed at the incredible prehistoric scenes before us. The island we were heading towards was large and surprisingly flat, covered in volcanic rock and cactus trees with no sign of any habitation whatsoever. We were so surprised to see so many brilliant white, sandy beaches as we cruised up the coast of the Isla de San Cristóbal, the fifth largest of the Galápagos archipelago. Heading into the only port on the island, we passed several small white cruise ships straining in the wind on their moorings. The bay held a collection of low-lying, nondescript pastel-coloured buildings nestled amongst the sparse greenery and black rocks. Colourful blue, yellow, and red little water taxis scooted to and from two substantial tourist jetties, linked by what looked like a boardwalk.

We dropped the anchor on the outskirts of the mooring buoys and waited for customs and immigration to clear us in. Well, that was an experience, I tell you. We think New Zealand Customs are strict! We had to wait a few hours for them as we had arrived earlier than scheduled. A water taxi arrived at our stern with not one or two officials, but *eight!!* We could not believe it; they just kept coming aboard. Imposing uniforms, big boots, guns...you name it. There were seven men and one lady. They inspected and photographed the fire extinguishers, the fridge (goodbye beef in the freezer, I'm sure they had a good feed that night), us, our papers. The click of cameras could be heard all over the boat as photos were being taken of our gear. The official customs doctor's eyes nearly fell out of his head when he examined our first aid kits; he was mesmerised! We probably had more drugs aboard than what the entire island of San Cristóbal had. He was most impressed. The friendly Spanish interpreter was a huge help and kept everyone calm. It was actually quite a daunting experience to have *Marmax* probed like this. She was our home, and suddenly we all felt on guard. None of the officials could speak English, but they were pleasant enough, though very stern and obviously took their jobs very seriously indeed. This was a point driven through to us time and time again over the next week of travelling between islands. Tourism here was serious business.

The fumigation guy arrived the following day with a *Ghost Buster*-size gun to fumigate the yacht. The silly thing was, we had only got a bee and a fly or two inside once we got into Galápagos; no insects had entered *Marmax* up to this point of the entire trip! Bruce was escorted ashore with our passports, and we were good to go. The yellow quarantine flag came down; we had officially arrived in Galápagos! We scored the good tick of health from the hull inspector, the doctor, the department of environment, customs and immigration, the safety dude, and whoever was participating.

You can well imagine how exciting it was to first visit an island in the Galápagos! We hailed one of the gaudy water taxis, clambered in, and took off to the nearest wharf, which had floating pontoons for easy access on and off the island. Eyes wide, we whooped with delight as we neared the shore! All over these pontoons, shiny sea lions lolled about in the sun, many of them with young calves suckling. Stepping warily over their bodies – at this early discovery stage for us, we did not know if one was going to rip our legs off as we crept past. Big sleepy eyes gazed up at us, a slow turn of a head, large fangs warning us to keep our distance. No one would want to mess with these girls! There were steel gates up the ramps to prevent them from roaming freely along the jetties. Enjoying their own natural tidal enclosures and fenced beaches, the entire village esplanade was fringed with these wondrous creatures. Their throaty barks became background music once you got used to them. Hundreds of females and their young pups also lounged in the sun on the main beach, closely guarded by a ferocious bull sea lion parading at the tideline.

Some smaller sea lions surfed happily on the incoming waves, swerving hard sideways and disappearing back into the deep water as the waves crashed upon the sand. Sitting in the waterfront restaurant, we were mesmerised by silly animal antics going on all around us. Slow, plodding, jet-black marine iguanas with curious faces, spiky dorsal scales, and their salt-encrusted heads sauntered along the sea wall. Brilliant orange Sally Lightfoot crabs scrambled energetically about the rocks just above the sea spray; there was always something going on. It did not take long before all the weird stuff from strange-looking animals became normal to us.

The weather was perfect; the temperature a cool 22–23 degrees even though we were right on the equator. The sun was brilliant, and the seas were kissed by gentle tropical breezes. We were blessed to be surrounded by beautiful islanders who had helped us no end and welcomed us with open hearts with those big smiles. The people of Galápagos were quiet and demure and mostly Ecuadorian; how fortunate and privileged we were to be here.

It had been no easy street getting to Galápagos, mainly because of our muck-up with booking late. There were not many overseas yachts in the country, as the cost of entry fees had become so prohibitive. $US3000 just for boat and crew. We would have probably spent this amount again on just being there, so start buying those lotto tickets if you want to visit! However, our agent had made the path smooth for our entry; the security was quite phenomenal. We had cameras being played on us no matter where we went. The shoreline, streets, jetty, and docks adorned with wild sea creatures mingled with visiting tourists. They were all being continuously monitored. You could not move from island to island without an agent. We planned to anchor *Marmax* up for the week in San Cristóbal and take fast ferry boat rides to the islands. It was only US$30 per trip each way and, with no wind, faster than sailing our yacht. Taking the taxi boat ashore was US$1.00 per trip, per person. We got to absolutely love this simple transport system.

Glenys had found a young boy of about 12 years old, Gabriel, to act as our travel agent while we were on land. The children of the island seemed to run many of the small businesses, including shops and restaurants. It appeared that, much of the time, the parents simply knew very little English. Little Gabriel was unbelievably efficient. His dad owned one of the fast tourist boats that supplemented their travel agency located in the main street. Gabriel ran around on his pushbike with his VHF handheld radio, leaping taxi boats and fussing around his clients. We could not recommend him highly enough; he did a stellar job for such a young man.

The travel agents of Galápagos were legally responsible for ensuring their clients got on and off boats at the correct times and at the right places. This sounds obvious, but the border controls between all the islands were tight as could be. We always had boat identification lanyards put over our heads before boarding the interconnecting island vessels. Bags were checked by customs, a personal minder was ever on hand to meet and greet us on each jetty, ensuring we got to our hotel or the next boat or tour. We had a roadblock one day on our trip to the highlands of Santa Cruz Island. All of our identification was checked, and we were asked if we had any fruit or vegetables on us, all for environmental control. The islands had different species, different habitats; they were understandably paranoid about insects and animal matter. Until recently, the future of humans there depended on the islands. Today, the future of the islands depends on animals.

At first, the constant scrutiny seemed very invasive to us, mind you. Once we were checked in and out a few times, it became a habit, so it was not a problem. The uniforms were quite intimidating too. Something so precious deserves to be heavily protected and respected.

The Galápagos Islands are an archipelago of thirteen significant islands plus a few minor islands. They are volcanic in origin and very geologically young. Basically, they are the tops of enormous submarine mountains and are dominated by lava flows, interesting basalt formations, and stunning, white sandy beaches; all very dramatic! The coastal areas and smaller islands were covered with bushes and small trees, mainly cacti, silhouetting the horizons. The islands have varying climates, mainly because it hardly ever rains there. Many islands are arid, though there is lush vegetation on some islands, where vegetables and fruit are grown. Most of the drinking water was distilled. Because the Galápagos lies in the meeting place of winds and ocean currents coming from the north and south, these determine the species' climate, distribution, and abundance. Each island is quite different.

We had arrived during what they seasonally call la garúa. This was the start of what the locals call the dry season, marked by cloudy skies, cooler temperatures, and misty rain on top of the hills. This runs from June to November, when the southwest trade winds are at their strongest. The unusual life in the Galápagos Islands is very much influenced by three ocean currents and the climate. The meeting of these three currents has led to the incredible marine ecosystem in Galápagos and the unique, often weird variety of wildlife found here. The most important is the Humboldt Current. It is bursting with nutrient goodness and also runs at its strongest between June and November. The Humboldt runs north from the freezing waters of Antarctica, up the western coast of South America, then swerves out to flow into Galápagos waters once it hits the equator. This also brings animals, which is how the penguins and fur seals first arrived in Galápagos.

The second current is known as the Panama Current. This comes down from the north bringing warm waters into the Galápagos Islands. When these two currents get together, la garúa begins misting up.

The Cromwell Current is a deep-sea undercurrent that runs from the west. Its job is to drive nutrients up from the lowest levels of the ocean floor up to the surface. The nutrients feed the plankton and plants, which form the basis of the food chain in the Galápagos. The entire ecosystem grows up from here. Charles Darwin was the great British naturalist who figured this all out.

According to locals, San Cristóbal was supposed to have had not a lot going for it. We loved it! To be greeted every single morning by playfully romping sea lions was very special. They blew bubbles against *Marmax*'s hull at night, we presumed because there were so many fish under us. It sounded like a game of peek-a-boo going on each evening as they looked at each other through the bow-thruster hole of the hull. We had to place fenders on the stern to stop the friendly creatures from jumping aboard; many local boats had lengths of barbed wire slung around their topsides to keep them off. A couple of unsuspecting catamarans had arrived. Their twin scoops on the stern were accessible platforms for the sea lions to climb. Imagine waking up in the morning to find your cockpit full of barking sea lions! How would you get rid of them? Flick a T-shirt in their face? The mess they made was not worth the cuteness! Honestly, it was just like sleeping on our farm during lambing season. The sound of the baby sea lions was just like lambs that had lost their mum. Of course, the mothers and fathers made sexy noises all night; we got used to it.

By day, we saw big frigatebirds take fish out of the sea lions' mouths as they surfaced with their catch. Frigates cannot dive in salt water, as it kills them, so they steal food. Often frigates can be found in the volcano lakes splashing about to get rid of salt from their wings. We had never learnt so much about animals in our lives. It was a marvellous education for Janelle, at her age. Each Galápagos animal had an incredible history and story behind it. The famous Charles Darwin Institute, among others, did a fabulous job interpreting it all.

We went on some extraordinary walks through the garúa mist to the top of volcanoes. The infrastructure of the walking tracks was astounding, all diamond-cut volcanic rocks flanked by incredible moon-like scenes of black rocks, lichens, and prickly pear cacti. We swam at the most beautiful beaches you could imagine and climbed high onto hillside rocks to sit down with blue-footed booby birds. We enjoyed animated conversations with young and cute marine iguanas and stepped warily around the big, wrinkly older fellows. They get uglier as they age. The iguanas were mainly black and grey, but the boys' skin turned on magnificent colours during the breeding season. Whoever was the most colourful apparently pulled the girls. Same with the boobies, actually; those with the bluest feet attracted the hot chicks! We visited a couple of the tortoise breeding farms where we met tortoises from only a month old, so cute! Then the grandaddies, some who reached an astonishing 125 years of age. The powerful jaws of those old boys crunching on thorny vegetation were phenomenal. It is hard to believe the first Spanish sailors, and then pirates, used to eat them! The tortoises could live for a long time without water, so they were kept alive on the ships and then given water only a short time before they were due to be eaten. The Charles Darwin Research Station's mission is to ensure they do not become extinct. From what we saw, the program has been hugely successful, with both small and super-sized tortoise critters looking very happy and robust in their bushy habitats.

Through stunning lava tubes we climbed; they were pretty different from our New Zealand volcanic environment. Lava flows on these islands had stopped rather suddenly, and because many of the islands are so young, the rocks were quite soft compared to ours.

We left San Cristóbal and caught a fast boat to travel to the town of Puerto Ayora on the island of Santa Cruz. The boat journey was the trip of madmen. Thirty people, primarily tourists, were crammed into super-charged cruise boats, travelling way too fast. Veering dangerously from side to side through foaming waves formed by converging ocean swells, we surfed off the reefs and the bouncing wake of other such boats. The crew seemed to enjoy the excitement and fear it created in the passengers' faces. It was like flying along inside a washing machine. It was a genuinely hilarious and exhilarating ride for our *Marmax* family. But if something was to go wrong, it would have gone really wrong! No safety drills…as for lifejackets? They were still in their plastic wraps, crammed into a cupboard up the front! There was a good supply of sick bags, and we did get a headcount going on and off, so we'll give them brownie points for that.

During a highlands tour, we checked out enormous sinkholes, giant tortoises, and an incredible treehouse that had accommodation both under the roots and up in the branches. Santa Cruz is the second-largest island in all of Galápagos. We were surprised to discover it had a population of twelve thousand people! It is the country's economic centre and has a bustling town occupied by dozens of restaurants, small motels, souvenir shops, convenience stores, and internet coffee shops. The main commercial jetty was a hive of activity, clearly geared up for tourists. It was highly amusing to watch the sea lions, hammerhead sharks, and golden rays swimming around the clear harbour, totally unfazed by all the action upon the water. Frigatebirds, pelicans, herons, and lava gulls soared about in the air, kind of like a surreal Jurassic Park scene!

For insurance requirements, Bruce had to stay back on *Marmax*, as we could not leave the boat unattended for longer than twenty-four hours. He took one of the wild, fast boat rides back to the yacht in San Cristóbal. Glenys, Janelle, Les, and I set off to the stunning island of Isabela. The streets were all soft, fine sand, just like castor sugar! By now, after so much island exploring, it was hard for us to pick a favourite island.

Wow…Isabela was a place that we could not get enough of. Brilliant, orange-coloured flamingos, thousands of marine iguanas, the Tintorera Grotto with whitetip reef sharks swimming with the sea lions. The sea lions nipped at the shark's tails to get their attention so they would play. We snorkelled with majestic turtles, rainbow-coloured parrotfish, feisty sharks, cute penguins, and those frolicking sea lions. We had a brilliant naturalist as our guide. During her fascinating narrative, she smiled as she swept her eyes across the scene before us and gently murmured, "We live in a paradise, within a paradise." Never a truer word said.

The animals posed like photographic models. One small rock protruded above the gentle tide. Here we counted one pelican, a booby bird, a penguin, a heron, dozens of Sally Lightfoot crabs, iguanas, and two sea lions. That, I repeat, was one small rock. The ever-present frigatebirds circled us from above. It was like an aquarium in heaven! Galápagos had a two-metre rule with animals, but it was the animals who broke this law. If you did not watch where you were walking, you were bound to trip over some magnificent creature. We enjoyed superb accommodation, sensational food, and drinks everywhere we went. I sang "Amazing Grace" in the local church; I don't believe my singing has ever sounded that good before; it just goes to show how good those pitch-floor church acoustics are!

Back to beautiful Santa Cruz for another night, and we walked through to Finch and Academy Bay along an absolutely out-of-this-world adventure walk. Through rocky trails fringed by cactus groves, salt lakes, mangroves full of awkward pelicans, powder-puff soft white beaches, stunning mudbrick homes, all leading to a beautiful cool canyon with a crystal-clear salt-water grotto to swim in. We were so excited to find this incredible place on earth, stunning Las Grietas grotto. It was quite an ethereal experience looking down upon its brilliant emerald waters embraced by golden rocky cliff faces. We'd brought along our snorkelling gear to explore the depths and came face to face with enormous colourful fish gently lingering below. The world of the Galápagos really is incredible.

Before this sailing trip, Janelle's Uncle Ian had given her a book called *My Father's Island*. It was a fascinating autobiography about a young girl called Joanna Angermeyer and her family, growing up in the Galápagos Islands from the 1930s. Glenys and I were into this historical stuff. We were absolutely rapt to not only find her Uncle Teppi, who owned Angermeyer Destinations in Santa Cruz but spend time with him. We enjoyed a final feast and cocktails at his establishment before heading back to San Cristóbal on the boat.

Returning to *Marmax*, we were bubbling over with excitement and stories of our adventures. Understandably, Bruce felt a bit left out, so Glenys re-booked the entire trip to Santa Cruz and Isabela! Bruce and Glenys went off on a second honeymoon for two days in paradise. In the meantime, Janelle was to stay with Les and me.

We decided to take Janelle for a swim with the sea lions who were basking in the sun on a beach off to *Marmax*'s port side. This was actually quite risky business. The bulls are so ferociously protective, and let's face it, we were on our own in unknown waters, alive with all sorts of creatures. Oh, my Lord, that water was freezing!!!!! Have you ever heard a fourteen-year-old girl squealing into a face mask underwater? The sea lions came rushing out of nowhere at us, as Janelle said, clutching her little heart, "I felt that their eyes looked right into my soul!" What an experience for us all; it sure got our adrenaline pumping!

Glenys and Bruce had an exciting adventure revisiting Santa Cruz and Isabela. They then headed back to re-join us aboard *Marmax* in San Cristóbal. We stayed in the bay for some final runs around the fabulous coastal volcanic walks. Sheer cliffs dropped off from the pathways into crystal clear ocean lagoons, playgrounds for our favourite sea lions.

One funny story here, remember the chapter about Virgin Gorda in the British Virgin Islands of the Caribbean? The story of us girls hanging up our clothesline with lingerie hanging up and a big tourist boat laden with tourists sliding in alongside? Well, while in the bay of San Cristóbal, this was only the second time we had that clothesline up with girly bits on it. Much to Bruce's disapproval, up it went.

Not thirty minutes later, guess what happened? A taxi boat turned up alongside with a wedding party aboard. Glenys was on deck, drying her hair with a towel over her head. They hollered out in Spanish; she thought they had asked her to take their photo from our yacht, of the taxi boat adorned with their wedding party. No, they did not just want their picture taken. They wanted it taken aboard *Marmax*!

Much to Glenys's surprise, the boat fronted, bow up, to our stern, stepping over our sleeping sea-lions on the tuck. A beautiful blushing bride and her handsome husband in naval regalia climbed aboard. Apparently, he was a ship's captain in the Ecuadorian navy. The resplendent bride tottered to the bow, with the help of her new husband, in ridiculously high heels and a magnificent, sheer, full-length wedding dress.

Glenys and I were both thinking, "Dear God, please don't let her rip that gorgeous dress off on the split pins of the shrouds of the mast!"

Janelle was frantically trying to pull back the clothesline tied up with four clove hitches to the rigging. Glenys got the giggles. Les was now tasked as the official photographer. Glenys had her hands full, apologising for the clothesline and trying to help Janelle. We were waiting for the bride to trip in her high heels, hole her lacey train, and catapult backwards into the water. It was simply hilarious. Seriously, the things that happened on this boat, you simply could not believe. It was wild!

Final swims with our gentle, whiskered friends. The last farewells to the friendly storekeepers who had helped us out. Farewells to the cafe owners who persevered with us, allowing us to spend hours on their wi-fi trying to get business done and blogs out. Farewells to our ever-smiling water taxi drivers and, most heart wrenching, the goodbyes to the frolicking sea lions of San Cristóbal. How we will forever miss their hilarious, watery antics around *Marmax*!

Clearing customs to leave the Galápagos was actually quite fun. We had to meet up with our customs agent, Carmel, at her dubious-looking office, three streets up from the waterfront. Her younger brother had bounded up the jetty to greet us in the dark, late the night before, to ensure we made our 10 am appointment the following day with her. We were definitely being continuously watched. I guess, if we were the keepers of these Galápagos jewels, we would do the same.

Up we fronted to Carmel's office, boat registration, clearance paperwork, and passports in hand. We were bundled into two taxi utes with escorts. At the time, we suspected these men may have been other relatives of Carmel's. We could not pay for customs, migration or yacht services in Galápagos without crispy US dollars in our hands. No Visa cards, no credit cards accepted. There was much to ponder and be suspicious about regarding the economic consequences of this loosely governed, cash society.

That taxi ride from the office was, let's say, an exciting jaunt, up and out of town, into the bush. Were we heading to the firing squad? It felt like it! Suddenly we zoomed left, up a rough, volcanic, scoria driveway. Here, we were emptied out at the steps of the department of immigration, a building strangely out of character with everything else Galápagos. A one-man show, who studied all of our papers again, eyeballed us. The usual secretive, soundless photo was taken of us by the customs lady, Carmel. Passports stamped, we were bundled up again, escorted back into the taxi utes, and dropped off at the customs office. Everyone shook hands, and we were free. We fully expected to be escorted again to the jetty and told to leave, as in Panama. This was an unexpected surprise and a much more pleasant way to leave a country. We never expected anything less from the lovely Galápagoans, who, besides customs, had only offered their smiles at every twist and turn we made across the islands.

A cracking 80-nautical-mile sail, against massive ocean currents and surf breakers miles out to sea, found me on my usual midnight watch. Glenys was on changeover with me, and suddenly this enormous blast of water whooshed out from our starboard side. Talk about give us a fright! We never saw the whale, but I spent the next hour expecting to slam into one! It was just past midnight and the dawn of Glenys's birthday.

About 20 nautical miles off Santa Maria, we realised we had developed severe engine battery problems heading out of Galápagos waters. Despite their ability to charge up, they were not holding their charge at all. This situation could seriously jeopardise the boat's safety, as we were heading to the land of hidden coral atolls. Besides that, our ability to make water and power and run the refrigeration and navigation instruments would be severely compromised. The inevitable decision was made to turn around and head back to the major commercial port of Santa Cruz.

We had mixed emotions as we headed back. Having already cleared customs in San Cristóbal, would the powers-to-be charge us another US$3000 to give us clearance back into Galápagos? The cash economy is a little scary when dealing with some of these foreign officials; they really do seem to have the power to do what they like. Glenys phoned our Galápagos yacht agent, Javier, from YachtGala again, so he could hopefully smooth the process for us and a translator needed, please. Under the authority of an emergency stop, we entered Academy Bay. It came as no surprise to hear one of the two lovely young guys from YachtGala was a nephew of Carmel, the official in San Cristóbal, where we had just left. As suspected, it was all in the family.

Bruce and Les snuck ashore that night to raid the island's ATMs for cash, as we could only take US$200 out at a time, three times, every twenty-four hours per card. We do not carry thousands of dollars while travelling through empty oceans. We had not re-cleared customs and immigration either, so it was a quick, fugitive-like visit!

During the next two days, we installed four new batteries. More speed dates with the beautiful scenery of Santa Cruz, some final ice creams for ice cream–addicted Janelle and Les, last-minute cocktails, another couple of hours wasted in the Santa Cruz immigration office with some excellent Spanish negotiating on our behalf by our YachtGala boys. We had to give exactly twenty-four hours' notice to gain clearance to leave Galápagos once we were sorted. Glenys had to go in the following morning to get that precious clearance again. Another customs inspection of *Marmax*, complete with an armed anti-narcotics man, and we were on our way. Thankfully, besides Bruce's surrender of many hundreds of cash dollars for batteries, marine electricians, customs fees, and translator and agent fees, we were not required to pay that extra US$3000. Smart thinking to use an agent in these waters!

It was weird saying a second goodbye. A final wave to the sea turtles, the sea lions, the incredible energy and colours of Santa Cruz. As the song goes, "We may never pass this way again…" The sight of Isabela's imposing volcano rising above the clouds in a farewell salute to us was eye-watering. We kept looking back for that final glimpse, a bit like saying a long goodbye to a loved one at the airport. Needless to say, Bruce was happy to leave the money pit!

Sadly, we had to move on. We were about to embark on a twenty-five-day passage to the Marquesas Islands. It was an honour and a privilege to have visited Galapagos and, again, so much more than we ever expected! There were so many stories within stories I could tell you about; our time there was simply magical.

First views of Galapagos—Kicker Rock

We LOVED the Iguanas!

...and friendly sea lions

13. The Pacific Passage: Galápagos to Marquesas

It felt kind of strange by now to be in our home paddock, so to speak, the South Pacific Ocean. New Zealand seemed only a few million stars away on a magical journey; we would never really, ever want to end…

Within the first forty-eight hours of leaving the Galápagos, we encountered a taste of the Doldrums, now referred to as ITCZ (how boring), the intertropical convergence zone. Feeling in the doldrums today? Fear not! You are not depressed; you are simply in the intertropical convergence zone!

The slapping of gasping sails on the mast and a sloppy, uneven sea drives you nuts when you endure it for days on end. The worry of stainless-steel bolts and screws working their way loose on mast and boom, the wearing of lines and sails, the preventer tight as a guitar string, stopping the jarring of the big boom. It is not fun, and it would be plain boring if not for the constant interplay of our aquatic friends, which kept us all amused as we sailed along.

The humpback whales seemed to be migrating towards the east. A whale submerged right alongside our yacht as we gently glided through the water. We must have given it a fright, almost running into it. A hundred metres behind the boat, it rose again with another whale alongside. During one twenty-four-hour stint, we spotted three pods of travelling humpbacks calling and blowing in the waves. Four enormous pods of missile-like dolphins shot straight up in the air and belly-flopped down beside us. You could almost hear them shrieking with laughter, mimicking us in equal delight. There were two different species. Small, compact dolphins and big, long barrel-like beauties turned upside down to show us their white bellies while racing our bow through the bubbling waters.

Millions of flying fish fired out like tracer bullets from *Marmax*'s waterline, some gliding over a hundred metres, in and out of the crests of the foaming waves. In the background of our cockpit conversations, we could hear the constant sound of *ssshhhing* as they hit the water. We also saw tuna schools bursting through the waves; shiny silver razorblades chasing bait. Tiny black birds darted about. We had no idea what sort of birds they were…There was no Google out there! When did they rest? Where did they go? Their constant chatter often sat in the dark skies above the yacht throughout the night.

The ocean was now changing from that Pacific emerald blue back to Sapphire gin bottle blue. Next time you reach for a bottle, imagine a little white yacht sailing along the top of the label. It was that blue. We thought we'd lose that colour in the Mediterranean; we also thought we'd lose it after the Caribbean. How can the beauty of this colour be described? We've been told the best is yet to come in French Polynesia and the Tuamotus. No wonder those six-month voyages by sailors often turned into six- to ten-year life adventures. It was clearly addictive.

Many people asked us how we kept boredom at bay; I looked at our videos. It was an amusing sight to look back on. We lifted weights while on watch, one- to two-kilo dumbbells. Les strapped them on the top of his feet to do leg lifts. This was my favourite exercise, especially with loud music in my earphones. Les was often found with his on, animatedly silently mouthing his Peter Frampton and John Denver songs. His hands moved to the beat in the air. Sometimes he forgot he was not supposed to be singing out loud; "Les! Shut up!!!!"

We tossed a basketball at each other in the cockpit. Boisterous pillow fights with Janelle, message-in-a-bottles biffed over the side. Glenys and I decided to tackle knitting, a talent I used to have as a teenager before life got busy. They say it's like riding a bike; I think not. It was difficult. We tried cable knitting...with three knitting needles. Haha. Glenys and I were in absolute hysterics. In the end, we had so many tears rolling down our faces, we couldn't see what we were doing anyway.

"Mummmm! We need you!" Glenys howled.

Our mother is a brilliant, talented knitter, but she was a long way away. We dissolved in a giggling heap on the cockpit seat; knitted pieces either finished with mysterious holes in odd places or unfinished and hidden in cupboards, out of sight, out of mind.

Hard to believe, but my champion fisherman on board, Mister Leslie Marsh, failed miserably to land five fabulous big mahi mahi over the next few days.

"C'mon Les, don't blame the fact the big gennaker is flying, and we are charging along at 8+ knots with a five-metre swell off the stern and can't stop!"

He had frustration written all over his face, and of course, we never let up teasing him. A medium size mahi was landed one night. We had it marinated in delicious coconut milk and lime juice for lunch. We should have invested in a fishing net as well as the gaff to get them aboard. It was extremely tough to get them safely on the deck from such a high stern on a constant lean; we all felt a bit sorry for Les. Happy to say, he did eventually redeem himself, and the fridge was filled again.

On the 25th of June, we were thrilled to receive a text, through the Iridium satellite phone, from my eldest son, Ben, and his lady, Katie, announcing the birth of their first child and our first grandchild. A healthy, bonny girl named Adalind. On a perfectly sunny South Pacific day, we celebrated with a champagne breakfast, then delicious pizzas made by Janelle and a fun sail with the big red and white gennaker billowing off the bow. We could not wait to meet her in October; we would not see her photo for several weeks.

One of our many amusing crew conversations was focused on distributing the final five precious bags of potato chips. At US$6-8 a bag in the Galápagos, we decided against such an investment at the time. Dumb. The options: would we just have a bag of potato chips for special occasions? (They might go mouldy waiting!) At our discretion, did we have one bag a week each Sunday (Glenys) or binge (my personal favourite) and have one bag each to feed out of? Big decisions were to be made here! The vote? One bag every Sunday at 5 pm. *Marmax* council was at a close.

Many of you would be surprised to hear it was quite cool on the equator. In fact, we had to forage around under bunks to find our jumpers; it was chilly! We certainly were not missing our New Zealand winter but hailing our blessed fortune at being able to sail the glorious South Pacific Ocean at this time. The countdown was on! Only three more months before we were back in Auckland. It was a confronting thought, a dream we'd all have to snap out of in a hurry.

My first overseas yacht trip was with my dad when I was sixteen years old. He and his crew had raced our family yacht up to Vila, and I was lucky enough to join him for the return voyage back to Auckland. I'll never forget the thrill. Ever since then, the electrifying excitement of being on the wheel in filthy weather, flying along with stinging spray and froth busting up around the boat, has always been the ultimate in intoxication to me. Maybe it's genetic; perhaps it's a touch of madness? Let me take the wheel when the weather wants a fight; I just love it.

I also realised on this trip that most people are on autopilot in their life. We all have the drumming of living life constantly beating in the background: family, friends, work, business, day-to-day chores, health, relentless mobile phones and emails…just life. Are we not all simply going through the motions rather than consciously living life? These were the thoughts that ran through my uncluttered, over-thinking mind while gazing with eyes wide open at the glorious night galaxies above. The masthead drew invisible lines in the skies, Glory, glory, hallelujah stuff. It was amazing. Words simply could not describe the profound beauty we were gliding under…and over.

Previously, I have mentioned the weird coincidental stuff that had happened to us during our sailing adventure. I was reading an article out loud in the cockpit one sunny afternoon. Glenys had found an incident report in the ship's resource and manual library, all about a 47-foot Moody yacht, which had collided with a tanker charging along at 27 knots in dense fog. Result: the boat sank, the tanker kept going, despite both being aware of each other. Mistakes were made by both captains in regards to their lack of understanding of maritime law. Compasses were set wrong and misread. There were many unfortunate and tragic events leading up to the big crunch. You learn something from anything you read. Let's face it…knowledge is power.

Having not seen a single boat, aeroplane, or human for two and a half weeks, I looked up from reading this article. There, swiftly closing in on us, was a large Chinese tuna boat on a collision course with *Marmax*. What do you think the chances were? Slim chance? Nope, not on *Marmax*! We could not pick up this boat on radar, though we had it within our sights as it came charging towards us. In no time, they were two hundred metres off our port bow. We were flying the gennaker, so we had limited manoeuvrability without dropping it. Was there anyone on the tuna boat helm? Neither of their two radars seemed in operation. Janelle zealously blew our foghorn, and tuna crew swarmed to the decks. While we were eyeing off the tuna vessel and quickly planning an emergency escape strategy, Les yells out,

"Fishing buoy to port!"

Yikes! The water was over four kilometres deep here. No radar reflector on this thing, a tiny bob in the ocean. Did it have a net connected to it? Not sure. We artfully dodged it as the friendly tuna crew aboard waved their greetings to us. We were still not sure if they were deliberately steaming towards *Marmax* because we were a curious sight. We were located 1700 nautical miles from anywhere. Were they picking up their buoy? Were we coincidentally on a collision course with it, then consequently them? We'll never know. It was simply another "What the?" moment and much too close for comfort! Pirates were always the first thing on our minds in situations like this. It can be scary out there in the middle of absolutely nowhere!

We ran into more trouble. The generator was cutting out when the water-maker or hot-water jug was put on. Constant chafing of the sails in the rolling swells had torn out the webbing and ripped out the clew at the foot of the mainsail. We would have to complete the voyage under a permanent first reef. The gennaker went up with no problem, but the detachable carbon fibre–covered bowsprit it hooked onto snapped in half under the load. Although we had daily checks on the gear for wear and tear, it seemed much could go undetected; stuff happened. We had to roll with the blows. I should never have said that. The next day, after I wrote those words, we did indeed blow out the gennaker. Shredded, you might say!

On a lighter note, there were a few trivial happenings over the following weeks; I know you love to be entertained!

Trivial happening #1: Being the night cook, I got to buy dinner ingredients when storing *Marmax*. We had finally got used to bright yellow chicken breasts in the Spanish countries. Apparently, the farmers fed the chickens corn, thus giving them a distinctive yellow colour. In Spanish, chicken is called pollo, so we had that name sorted. Chicken meat was a primary staple food in most countries we had visited so far; it was right up there with seafood. At this stage of our Pacific Ocean passage, we were getting down to the very last of our fresh meat and vegetables. We carefully studied everything we bought, as we did not understand any other languages besides English. Most writing on the side of food items seemed to be, what we thought was, Spanish.

When in Galápagos, I came across what I thought were chicken breasts. Upon opening, no…it was a whole chicken, stuffed full of not only its own feet but his buddy's as well. Yes, four feet. Very well disguised for unsuspecting foreign travellers! They were sewn up inside; we could not believe their feet were so big! At least in Galápagos, their bodies were a healthy pallor! Just what were we supposed to do with all those legs?

Trivial happening #2: One night, I delved into the freezer to get another bag of pollo out. I was looking forward to cooking chicken for the crew, as we had enjoyed a nice run of fish for quite some time by then. The commercial plastic bag said it was pollo. It looked like cubed chicken, smelt like cubed chicken. You know what? We still don't know what it was. Bruce thought perhaps it was lamb kidneys. The question arose as to whether or not it was chicken kidneys or giblets? Could it have been veal? We are farm girls. This meat did not look like anything I had ever seen! Anyway, it was apricot chicken/ lamb or veal or something on rice that night, and I was rewarded with a withering look from Janelle. No one died overnight. It tasted pretty good, I thought!

Trivial happening #3: Back in Colón, Panama, we obviously stocked up the boat. One of my purchases was three months' worth of chewing gum. Great to keep you awake if you are trying to read a book on a rock-a-bye-baby boat in a swinging ocean for seemingly endless days and nights. So I'm into my stash of chewy one night on watch, a deadpan, glossy sea, smooth as a swilling glass of wine. Noddy stuff. Pitch-black night, a brilliant Milky Way above, a long watch ahead. Hand into my stash, fumble in darkness with the packaging, popped it into my mouth…Holey moley! I appear to have bought three months' worth of throat lozenges. They nearly blew my head off!

Trivial happening #4: The coffee saga! We bought a gorgeous big drum of Colombian coffee in Tortola, the BVIs. Thought it was instant, looked instant, wasn't instant! Much to our dismay, and having opened it after weeks at sea, we discovered it was plunger coffee. There was no plunger on board, so ever-creative Janelle sewed us some little reusable cotton tea bags out of an old pillowcase. Bless her. Bob's your uncle! Just a right royal pain when the yacht is throwing you about at sea. Thank goodness, the weirdest "got everything" shop in San Cristóbal, Galápagos, had a red coffee plunger waving at us from the shelf. We just had to be careful not to smash it!

Trivial happening #5: The argan oil saga. Most women know that Moroccan argan oil is the so-called wonder product for both hair and skin. Thanks to the generosity of our Barcelona superyacht team, we had several bottles of this wonder product delivered to the good ship *Marmax* at the outset of our big journey. So one morning, I'm conducting English lessons with Janelle at *Marmax* College, and she says to me, "Deba [yes, Deba]…what is that on your face?"

"Oh, it's that lovely argan oil stuff the girls sent over; it's perfect for your skin!"

Janelle replies, "Isn't that stuff shampoo?"

Hmm…once more, the label was in Spanish. Glenys had been using it as a body wash, dish detergent, and emergency laundry detergent. I had used it as a moisturiser. Actually, again, it could have been anything really; we had quite a few bottles to get through. Our translator app called it a "bus stop"!!

Trivial happening #6: In Colón, Panama, you may remember we shopped at the massive duty-free city, the Free Zone. I found this lovely men's cologne for Les for an unbelievable US$2 per bottle. He was happy to use it on the boat and gave his fancy stuff a rest.

"Have a smell of this, Glenys! It's gorgeous!" I yelled as sisters do.

So, Glenys bought a bottle. Now we were at sea; we had discovered that up fore we had Les wafting through the yacht with a very sexy scent following him. What did Glenys do with hers? She thought it was a really lovely smelling toilet freshener. Yep, you guessed it: eau de toilette, which apparently meant toilet to Glenys. Bless her!

Trivial happening #7: So, we had Glenys and Bruce's lovely looking fourteen- nearly fifteen-year-old daughter, Janelle, obviously travelling with us. Since the start of the journey, the boys had a beer (or three), and Glenys and I partook in cocktails whenever we were in port. We often ordered a similar cocktail for Janelle as a treat. No, not with alcohol.

Glenys requested a virgin cocktail, usually a piña colada. Now, previous to this trip away, Les and I had never heard of this term of "virgin drinks". (Glenys insists everyone knew it meant a cocktail without alcohol). When she ordered, it often left me cowering behind the drinks menu. Use your imagination, folks. We were usually served by young male Latino waiters, most of whom knew very little English. Here was Glenys gesturing her hands towards Janelle, stating her request. "A virgin cocktail please?"

You've heard the term "eyes as round as saucers", or "his eyes lit up!" Clearly, these boys know a few English words but seemed confused about what was on offer or what was being asked of them! We had to change Glenys's language. I believe we left, in our wake, a long line of wishful puppies.

Man alive, the stories we had from Galápagos! We bought weird root vegetables that we could not even pronounce their names. We purchased bags of what we thought was pearl rice but was tapioca. We bought peanut butter out of twenty-gallon drums. We watched a supermarket butcher drag a massive, dead, skinned cow across the concrete through clear plastic doors and hack it up with a machete. He then bagged it in huge lumps for the supermarket fridges. Delicious for Jamaican curries and stews. The sight would have been enough to send anyone but two hardened farmers screaming out the front door! Cooking on the Pacific passage brought many a smile to our faces, even before we cooked it! Oh, the memories!

We were down to the last few days before we reached Nuka Hiva in the Marquesas. So exciting! The cooking was getting more creative by the minute; we were down to our last white onion. Thank goodness the fridge was cold, and this time we did not run out of grog. Janelle was now scoring A's and B's in the majority of her schoolwork. Trying to get her to read books was a big challenge. She had handled the voyage, since day 1, incredibly well. As many of you will know, fourteen-year-olds can be pretty damn scary in closed spaces. Then again, imagine being fourteen and putting up with us lot in such a confined space. Go Janelle! We were all in good spirits and healthy as buck rabbits. This passage had taken us a long twenty-three days. *Marmax* had slowed a little with a forced slab in the mainsail, but hey…better late than never!

Putting up the spinnaker while crossing the Pacific

Marmax College at work

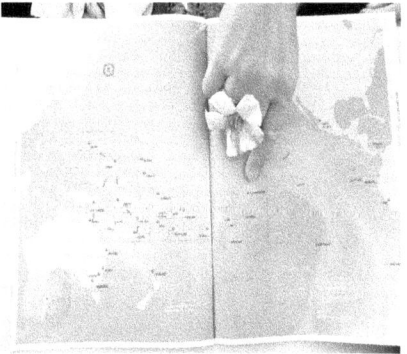

The best place to sleep in rough weather

Debbie & Les's sleeping quarters

14. The Magical Marquesas!

We had arrived in inky darkness at 4.30 am in Taioha'e Bay, on the island of Nuku Hiva in the Marquesas Islands of French Polynesia. One of our fellow Sandspit Yacht Club members, Ross Sutherland aboard *Anatere*, travelled a couple of weeks ahead of us from Panama. He had confirmed for us that Taioha'e was a good, safe anchorage for a night-time approach. We had been warned to be careful in approaching in complete darkness. The anchor lights of moored vessels could easily be confused with streetlights ashore; that was assuming every boat had its anchor lights on! There were indeed thirty-four boats inside the bay. To play it safe, we anchored on the outskirts of the fleet as the sun came up from behind those dramatically beautiful soaring mountains that surrounded the entire bay. Wow! We couldn't tear our eyes away from it.

I asked the *Marmax* crew for three words each to describe their first impressions of the Marquesas. These were their descriptive words: dramatic, spectacular, non-touristy, majestic, untouched wilderness and beauty, pristine mountains, beaches and shorelines, stunning. You get the picture. At this stage, we had not even gone ashore! I was almost at a loss for words, a situation many will find impossible to believe.

Many of the other travelling yachts anchored in the bay had their families aboard. It was such an inspiration to see so many grey nomads about too, so happy, so tanned, fit, and vibrant! We had to move the boat, looked over the side in total dismay…poor *Marmax*! The filth up the hull was unbelievable! Those alien barnacle things we had first seen in the British Virgin Islands had struck again. Not one of us aboard was brave enough to take the plunge and clean her up. Gigantic manta rays, oversized hammerheads, and other precarious-looking dark sharks were swimming about the bay. Frantic thoughts of being eaten by these ominous beasts swirled through our heads…

The unbelievable beauty just got more and more impressive and, by now, had become part of a typical day in paradise for us. While we had mentioned the word "amazing" so many times on our sailing adventure, the overused term "unbelievable" was now dancing on our lips. We were staring open-jawed at every glorious bay we ventured into. If you could drink with your eyes, we would have been legless by the time we left this kingdom of paradise.

All were eager to get ashore after our twenty-three-day ocean passage from the Galápagos; the first challenge was that the outboard motor would not start. Having been fully serviced in Barcelona and hardly used, surely this problem was going to be a simple one? After an hour of trying to get it going, we resigned to rowing, as we were booked to meet up with the clearance guys at 8.30 am ashore; time was ticking on. With Glenys and Bruce on auxiliary kayak paddles up fore and Les rowing, we set off. It was a long row ashore in a heaving swell. Folks don't call this place Rolly Bay for nothing!

With the usual raucous behaviour from our family, we sang "Row, Row, Row Your Boat" with Les in command. Much to our surprise and utter relief, a friendly Kiwi couple came to our rescue and towed us in. Ian McGowan and Kate Moss aboard *Allesea* had been travelling for a couple of years. They had bought a beautiful Hanse 50 yacht in Sardinia and were on their way back to New Zealand with their twelve-year-old daughter Millie. These guys were the first of many friendly floating Kiwis we were to meet in French Polynesia. People who spoke our language – hooray!

Our anchorage in Taioha'e Bay was formed from a blown-out volcano crater from ancient times. The sea rolled in and spent its energy smashing ashore, rolling the guts out of every yacht and catamaran in the bay as it swept by. As a consequence, landing ashore was quite a mission. We had no doubt there must have been many an unfortunate mishap here. We rigged our inflatable tender with two of our large fenders tied to the gunnels. This served us well, as the concrete walls of the dock were covered in sharp oyster shells. Other tenders and boats (all creatures great and small) lurched and collided with one another while tied to the gnarly wall. You had to wait until the next rolling wave came in and, at the height of it, make a leap for land. Practised and nimble, Janelle had no problem with this gymnastic requirement. In the early mornings, the local fishermen gutted their fish off that same wall; the enormous sharks that fed ravenously on the fishy feasts were frightening, to say the least. One would not want to fall in. The locals delighted in telling us the sharks were so well fed here, they would not eat you. We were to err on caution on that one, clinging to our darkness of doubt.

After a return tow out to *Marmax* by a lovely American guy, Bruce soon sorted the outboard. It was merely the carburettor float sticking, and he tapped it out with a screwdriver. Crisis averted!

The Marquesas are a group of islands forming the most northern territory of French Polynesia and are made up of six large and six small islands. The main six islands support a current population of around nine thousand. During the 18th century, the population was estimated to be at a whopping sixty thousand people, predominantly from Martinique and China. Others from Asia were brought in to work various plantations. Due to copious and erratic rainfalls, plus prolonged droughts, the plantations failed, and the labour force returned home. The islands are all mountainous with jagged profiles of vertical cliffs cutting into deep, tropical, fertile valleys, where most islanders live. Interestingly, the volcanic islands have no coral reefs protecting them outside, so navigation was easy for us mere mortals cruising about.

French Polynesia is a territory of France and is made up of five archipelagos: the Marquesas, Tuamotus, Societies, Gambiers, and Australs. Oh, to have the time to explore the entire lot! It was so far from home. It came as no surprise to hear of Kiwis leaving their boats on the hard up here for nine months so they could enjoy three months of each year exploring the place.

Back to Nuku Hiva, what a paradise! As soon as we hit terra firma, we found wi-fi. I was, of course, extremely excited to finally meet my new granddaughter on my phone! The farm at home was in the middle of a busy lambing season; our pet sheep, Millie, had also had triplets. Of course, all of us had a month of catching up with our world.

It took us three days to clear customs, as we had arrived on a Saturday; Sunday was Bastille Day, and Monday…well, they simply did not turn up at work. Were they hungover? Such is the island way. We replenished our fresh fruit and vegetable stores. Met lots of fascinating world-travelling yachties just opposite the dock, where wi-fi beamed out from a friendly cafe tent and the yacht agent's office. No sitting in the dirt here while you got your online business done. Tables and chairs – bliss! The islands were immaculate as far as rubbish and the care of the townships and harbours were concerned. Though in Nuku Hiva, there were mangey, friendly, dogs everywhere. The heat of the land was a bit hard for us to handle at first, with hot sun and regular tropical downpours. This explained why every island was so deliciously lush, plentiful, and green. Quite the opposite to the hurricane-whipped BVIs and the moonscape of the Galápagos Islands.

Much to our unexpected delight, we were lucky enough to arrive the day before Bastille Day. After seeing only miles of ocean for so long, the vibrant colours, energy, pomp, hilarity, and laughter of the people was a sheer delight and quite bedazzling for us. Stunning floats, enthusiastic maidens, grinning children, fearsome warriors on horseback, hakas, flowers, coconut-stripping contests, traditional singing and dance competitions, markets, gorgeous flower leis, and colourful pareos of the women. I personally fell in love with the huge skin drums with their throaty, booming beats. If we had the room up fore on *Marmax*, you would have spotted me marching off down the road with my arms around one of those big beasts, heading off out to the boat!

The islands use a French Polynesian currency, pretty, colourful notes with prices for goods close to New Zealand; apart from alcohol – oh my Lord, that was expensive! We dined in style at Pearl Lodge, welcomed by a truly gorgeous waiter called Jean Luke. Beautiful golden skin, perfectly shaped eyebrows, a very sexy shimmy, a halo of flowers around his head, and a smile to light the world up. He decorated our cocktails like we were all royalty; what an asset to that business!

The following day, we hired a 4WD Toyota dual cab. Glenys, ever the travel guide, gleaned a ton of information off the lovely tourism centre lady, Colette, and off we went. It took us a few shots at finding the road out of town; all roads led to nowhere, it seemed. Up the steep mountain passes, the views were absolutely incredible! Sheer cliffs to our left and postcard-perfect scenery to our right. This is a copy of Glenys's diary this day:

"Early start. Got rental at 8.30 am. Headed off around the island in a 4x4 Toyota. Awesome! Rained on and off all day. Went around the entire island, and only fourteen litres of fuel were used all day. FANTASTIC. Very rugged driving with Les at the wheel. Later learned we were "off-limits". No one goes there without a chainsaw, chains, spades, etc. Moonscapes, desert, bushwalking, drove past weed-eater and bulldozer men keeping the tame roads clean and tidy in the woop-woops. Past amazing villages on the rivers, wild horses on the roads, many with ropes around their necks with foals afoot – skinny as. Lots of pigs and piglets + roosters. Ruins. Coffee at some little place in Ha'atuatua. Beautiful beaches. Coconuts, bamboos, layered trees, very, very tall mountains, stunning concrete roads to goat tracks, but passable – just!"

We had an absolute ball with much laughter despite the sometimes utterly dangerous and precipitous wilderness tracks we were driving around. It was just beautiful immersing ourselves in a culture that had largely defied modern living. They were clearly succeeding in leading a simple life of self-sustainability as they had for the past hundreds of years. The car rental people must have gotten wind about our adventure. After hearing of our escapades, a Kiwi couple tried to hire a 4WD to make the same trip. The car hire people insisted they go with a guide because some crazy Kiwis had just gotten away with murder. Crazy Kiwis? Us?

We had a few days of R&R and the arduous job of manually refuelling the boat with four hundred litres of duty-free diesel with orange plastic jerry cans. This meant borrowing a ute, several trips to the fuel depot in another bay, then transfers in the tossing tender to ferry out to *Marmax*. It was quite satisfying for us girls to watch our two men doing some tough work for a change!

Later that day, we joined a few of our new Kiwi pals and took off for Taioa Bay, aka Daniels Bay. High cliffs with spectacular jagged peaks, lush tropical palms, coconuts, and fruit trees fringing freshwater rivers. The bay was alive with enormous manta rays. At that time, we thought this was the most beautiful bay we had ever seen in our lives; since then, we have seen such beauty in other places. We shall argue forever, "What was the most beautiful?" Some of these bays did not even look real. Blink, blink…no, they are still there. Wow!

Two hundred years or more ago, there was a substantial population in the valley here. The evidence was in the many ruins we explored going up the valley to a waterfall, high up in the cleft of a mountain. There were now only three families left in the bay. We joined three other Kiwi crews for a two-hour tramp up through the most stunning, tropical scenery imaginable. Navigating through swift-running rivers, lush forests, and imposing derelict ruins, our faces constantly upturned to take in the surrounding beauty of what was a true paradise. There were many impressive stone foundations, called pae pae hiamoe. The bases of ancient thatched houses had been seated on them all those years ago.

On the way back to our boats, we were treated to a traditional feast of fresh BBQ tuna, luscious pawpaw, mango and lime juice salads, with double-fried breadfruit wedges, deliciously washed down with cool pamplemousse, lime, and water. Pamplemousse are a Pacific grapefruit, wickedly addictive, so sweet and juicy! On this occasion, the juice proved to be somewhat fatal to many of our landing parties. Some of us went down with Bali-belly, or a gastro bug. Was it our host's water supply? After buying loads of fruit from this friendly, hospitable native family, we happily hauled it all back to the beach, where our tenders were tied up to coconut trees. This was yet another wonderful life experience for us all. We capped off with a night on the turps aboard one of the catamarans. Let me tell you; we slept very well every night in the Marquesas.

After spending three days cleaning *Marmax*'s hull of those alien-like barnacles, we headed back to Rolly Bay, picked up two of our refilled gas bottles, blasted off a few more emails, and restocked with groceries and alcohol. Wishing for no stone to be unturned, we then took off for a sailing circumnavigation of the entire island of Nuku Hiva. We stopped for some superb snorkelling along the way, arriving at sunset at an absolute stunner of a place called Anaho. Another big WOW! Another candidate for the most beautiful bay in the world, this time, endorsed by many before us.

Isolation protects Anaho, as it is preserved by its inaccessibility. Visitors must either walk here from Ha'atuatua, through a very steep forest walk or arrive by water, as we had. Over eight hundred years of Marquesan history is preserved in Anaho. Besides the massive cliff faces surrounding the bay, Anaho's reef, which swings in an arc following the beach, is one of Marquesas's most extensive reefs. Coral reefs are rare here, and it was strange to see them inside a bay, not protecting the outside of the islands as in the rest of the Pacific. We snorkelled for several delightful hours, with *Marmax* safely anchored just a few metres off it. What a day!

'Ua Pou is the third largest island in the Marquesas and drop-dead gorgeous, even from miles away. Incredible jagged pinnacle rocks soared up high into the clouds in the sky, rising from a carpet of green valleys and lush wilderness. There was more hand-on-heart stuff for our family. We dropped the anchor for lunch then wandered around the rugged yet majestic coast. Tiny native villages could be seen along the coast; quaint churches and steep, technicoloured cliffs kissed the ocean below. Fertile valleys with mango and coconut groves patch-worked the hillsides, just heavenly. There are indeed few things better in the world than to sail along an unbelievably stunning coast on a yacht in perfect weather. Sun shining, music playing, wind in your hair...simply divine!

The next night we pulled in alongside an Israeli yacht on anchor. The following day, we explored the pink rocks, sea caves, more awesome vertical cliffs, and a traditional fishing hut on the hill overlooking the bay. It was so much fun exploring these places in the middle of nowhere. There were lots of inquisitive dolphins curiously gliding around *Marmax* while we swam.

With the weather starting to deteriorate, we found a rather exposed anchorage outside a village. Strong winds were building and started whirling around massive columns of rocks. It was too rough to get ashore, so after dinner, we set off for a night sail to Hiva Oa, the next island south of the Marquesas Archipelago. Sadly, the wind and tide were severely against us, and we were going nowhere fast.

We never made it to Hiva Oa, forced to change course for the island of Tahuata. Hanamoenoa Bay was known as the only white sandy beach in the whole of the Marquesas, and what a beauty she was! Loads of manta rays hung out on the parameter of the anchorage, giving us and the six other foreign yachts plenty of entertainment. We arrived during the night to find the water alive with quite sizable fish attracted by our spreader lights. Much to our horror, the water was also awash with deadly box jellyfish too! We had been hanging out to go swimming!

In the morning, we sent Bruce, with mask and snorkel, over the side to test the water, so to speak. The coast was clear; no jellyfish in sight, so we all jumped in like a bunch of kids. We witnessed the most glorious sunset, complete with a yacht sailing through the middle of it, during sundowner drinks. The famous green flash, rarely seen in the tropics, glinted at us as the sun sank below the horizon. Perfection.

Arriving in these waters as the sun was setting was magical. The bays were stunningly B-E-A-U-T-I-F-U-L! Les pulled in a gorgeous yellowfin tuna for dinner, which we feasted on that night.

Towering cliffs were a backdrop for a coconut-tree-clad mountain that ran all the way to a fringed boulder shoreline with one little hut in the middle. Under those boulders, brilliant blue waters opened a treasure box of tropical fish and corals. Les and I were absolutely thrilled with massive schools of yellow and golden fish ebbing in and out with us in the tide. The volumes of what was there was remarkable. People had told us along the way that the Tuamotus had the absolute best diving in all the world; this is where we were heading next. We could not wait!

The second marvel of this perfect bay was that it was a well-known dolphin sanctuary. They welcomed us when we arrived in the setting sun, they saluted us as we sailed out, but the most incredible thrill of all was swimming with them during the day. Bruce was first over the side, then Janelle. Pretty soon, Glenys and I joined them, with Les spotting for us. Magic times. Glenys was beating the drum for English and maths lessons to recommence. Seriously, there was so much action going on, not even the teachers could concentrate.

Was this the most beautiful place in the world? We were off for our final sailing passage to the most southern island of the Marquesas, Fatu Hiva. The exciting thing was the bay we were heading to,

Hanavave Bay, had already been well documented as THE most beautiful bay in the land.

You will have to imagine the expressions on our faces, enthralled with an incredible double rainbow display that welcomed us into Hanavave Bay in Fatu Hiva. We sailed right under this ethereal arch of colour in a gentle, misty rain, and it all opened up for us as the sun was slipping gently away on the horizon. The rain lifted and revealed another glory, glory, hallelujah moment. What a diamond this place was! Was it real? Was it a movie set? *Planet of the Apes? Avatar? Jurassic Park? Gorillas in the Mist?* Wow, wow. WOW!! I took a thousand photos of this place! You would have to see it to believe it. We were the only boat in the entire bay.

This idyllic little harbour was also known as the Bay of Virgins. The distinct image of a virgin sat upon a mountain pinnacle, presiding over the village. She was pretty impressive! The volcanic upheaval that had created these oceanic jewels must have really been tremendous. What they had created was indeed out of this world!

At 8.30 am the following day, we were off to explore. The main street branched out into around five others. The entire island population was approximately nine hundred, with just over half living in Hanavave Village, half sixteen kilometres over the hill in Omo'a. Wandering through the township gave us the feeling of what it might feel like to walk through the Garden of Eden. If a seed had been dropped on the ground, I am sure it would have sprouted forth and begun growing before our very eyes.

There were many foreign influences in the various designs of houses. A Caribbean-like palette of paint colours dominated: bright blues, greens, pinks and yellows. Hibiscuses, frangipanis, tree orchids, and canna lilies exploded from between voluptuous stands of bananas. Fruit trees laden with starfruit and plump mangoes dripped in an orange glow. Breadfruit, limes sprawling and squashed all over the roadways. Pamplemousses heavily weighed down, bending branches outside the buildings. As usual, upon hearing we were from Nouvelle Zélande, the locals eagerly wanted to rejoice and show their knowledge and love of our All Blacks and the Maori haka. (Apparently, there is some fascination with the length of our New Zealand tongues: new to me!) Many Marquesans begged us to speak to them in Maori, which was kind of embarrassing!

Clearly, New Zealand was well respected in these islands. Glenys spoke to a native lady who had been to New Zealand twice in the past couple of years for a much-needed heart operation. They seemed ever so grateful and made us most welcome to roam about their island freely. They were also fiercely proud of their own heritage. We chatted with one young man who jubilantly threw his arms up, declaring that his Fatu Hiva was a paradise. No arguments there, mate!

We followed a wide winding concrete roadway up the hill, under awe-inspiring, steep black volcanic cliff faces surrounding the township. We were attempting to find the local waterfall, which a few yachties had highly recommended we visit.

"Look for the little stacked rocks…thirty minutes that way." A grinning young man pointed.

Hmmm…we finally found the little stacked rocks to show the entrance of the waterfall path. Trudging up a precipitous mountain into the clouds is not exactly my favourite form of a fun sport. But Glenys, in her never-ending desire to climb the highest mountain she can find as soon as she arrives in a bay, could not be stopped. We left Les and Janelle, prostrate and expired from the heat, on the side of the road, under a mango tree with a shrine to the Virgin Mary watching over them. Bruce, Glenys, and I soldiered on to the top for another hour. Let me tell you, I was still suffering five days later with ruined calves – I was wearing jandals. Les had my all-terrain joggers in his backpack at the bottom of the hill. Gorging on sweet mangoes along the way, the agonising climb was well worth the ordeal; the views were absolutely incredible! *Marmax* sat majestically, like a rare jewel, in the middle of the bay; it felt kind of strange to be so far from her up here; she was, after all, our one-way ticket home!

The tops of the mountains were flush with swan plants swaying about in the wind. Totally volcanic terrain with fertile valleys bursting with wild mangoes and coconuts, all the way down to the ocean. Melodious, colourful parrots flew below us; even the mountain goats were feeding below us; we were up in the sky! We headed off to collect the young and old from the roadway miles back down the valley. With leg muscles on fire, we continued down the mountain in an attempt to find this elusive waterfall that we had obviously missed on the way up.

There were no signs of the waterfall whatsoever. We took a wild guess and walked up what looked like someone's private muddy driveway then into acres of green Singapore daisy carpets. They were so green; it was almost blinding. I picked up a few human foot and horse hoof prints. This had to be the track. Over clear running streams, we finally found the entrance to a pathway flanked by those little stacks of stones like Hansel and Gretel crumbs. The path went vertically up, curving around moss-covered boulders. Thick vines like giant's arms wound around the trunks of ancient drooping trees. The oppressive humidity was exhausting. Then, like all promising pathways end, we burst upon the waterfall, a mirage of cold, sparkling water with tiny waterfall diamonds falling upon the surface. It chilled us to our core, but oh, so wonderful! Completely refreshed and reinvigorated, we revelled in an easy but slippery descent back to the wonderworld below. It was all quite surreal. "Can you believe this?" I kept hearing. Thank you, universe, for leading us to this paradise.

We had managed to have this incredible wilderness experience before the 2.30 pm supply boat arrived, lucky timing! The *Aranui* had arrived with three hundred passengers aboard. We had also dodged a significant rainstorm that had closed in ten minutes after we returned to *Marmax*. We need not have felt sorry for those three hundred passengers who could not view the entrance's beauty because of the rain. Like us, the rains drew aside, the light shone through the valleys behind the Virgin Mary rock, and Fatu Hiva's beauty was once more revealed like a stage curtain had been opened. While we had a late lunch in the cockpit, we waved at the camera-snapping passengers, ooing and ahhing on their way in, tourists on people-mover barges. We were so happy they, too, could experience that tremendous rush that we had while approaching this incredible island.

That very dark night, after yet another spectacular sunset lit up the cockpit drink time, another foreign yacht made her way into the small harbour. We watched her lights come in from far away and estimated her to be around fifty feet long. Sails down, they dropped their anchor, and the dingy was hoisted over the side from the top deck at lightning speed. There seemed an urgency and panic in the crew's tone, foreign accents we could not decipher. Gee, we hoped it was not a mercy dash. Was this an emergency health problem? A fight amongst the crew? Crew getting kicked off ashore? All seemed well in the morning as the yacht sat serenely at anchor, all quiet. Quite puzzling.

The following day, we set off around the corner to the township of Omoa. We had heard this place had a supermarket, infirmary, and museum and generally housed a more significant part of the population. This sounded interesting.

A rolling surf met us, crashing onto the rocky beach. We dropped the anchor anyway, putting out a ton of chain for safety. Into the tender, we braved the swells and ducked in to tie up behind a slippery landing dock. We set off on one of our exploratory visits. Golly, it was so humid here! This stunning little village was surrounded by volcanic rocks with extremely fertile hillside coverings. Native trees, flowers, fruit trees, and crystal-clear streams all provided ideal glasshouse conditions. Phew!

Les and Janelle, the ice cream addicts, strode off to find a shop that sold them. We found a building. Les, was clearly excited. He had checked out the enticing advertising signs, all in French, and worked out which delight he was going to buy when the place was due to open in the afternoon.

Janelle piped up… "Er, Les? I don't think they are ice creams. They look too healthy."

Crushing disappointment! I do believe we had found the infirmary! It was a poster offering healthy alternatives to the junk food our crew was craving. Funny as! Two hours later, the frustrated duo managed to score chocolate Magnum ice creams at the supermarket. It was an anxious, nail-biting two-hour wait for the shop to open. We later heard the crew we were about to meet at the supermarket had also been similarly hoodwinked! It was pretty good marketing…it got Les interested!

Walking down the main street, I turned to look back at the waterfront. Oh my God! Had *Marmax* dragged anchor? She looked like she was sitting right on top of the surf line from where we were standing. "Glenys! Look at *Marmax*! Has she dragged, or did I see things?"

Glenys sped off at a full run back to the waterfront; she, too, freaked out. Thank goodness it was an illusion. We had anchored so far out, so how could she look so close?

Glenys, thankfully, gave the thumbs up from five hundred metres away. All was well. Another crisis averted.

After meeting dozens of chirpy little chickens, bantams, puppies, and goats all over the town, we also found time to rescue a hungry, thirsty, foot-fettered, tangled-up black pig. We walked as far as we could while killing time for the long-anticipated 2 pm opening of the little supermarket. While waiting out the front under a tree, along came a couple of yachties. World-wandering yachties appear to have a great affection for Croc shoes. I wish I could have photographed them all; they come in hundreds of versions! These two guys both had pretty impressively coloured Crocs. Curiously, they were swinging a giant shopping bag around and around in the air; was this nervous energy? Yes, they were yachties from Ukraine. They were an absolute hoot! Turns out they were off the yacht that had arrived in the middle of the night from our previous bay. The inevitable question was raised by the interviewers, Glenys and me.

"What the heck were you guys up to in the middle of the night??"

Giant grins on their faces and in broken English… "Well, we were out of alcohol!"

"What…? But the shop was closed," we exclaimed.

"We went and knocked on all the house doors, and a little man sold us some homemade cider!"

"Wow, you guys were desperate! Let us guess, have you just arrived straight from the Galápagos?"

"Yes! One week and we ran out of whiskey and beer. No wine, no rum…nothing. We were definitely desperate!"

We found this highly amusing, as we had also had the same problem aboard *Marmax*, as well as no coffee. Clearly, the *Marmax* crew were not so reliant on alcohol!

Now that would be a fun ship to sail! These guys were an all-male crew of three, all in their forties and on their way around the world.

"Where are you heading to?"

"Around the world."

"Where is your next stop?"

"Not sure."

"How long are you away for?"

"As long as it takes."

Unbelievably common questions, with common answers from our watery friends. And again, as usual, when they heard we were sailing from Spain to New Zealand in seven months, inevitably their mouths dropped and their eyes opened with surprise…

"Why so fast??!!"

Yep, call us foolish!

Another potential crisis was put to bed. Just before the supermarket opened, a native lady pulled alongside us in her ute with a panicked look on her face, asking if any of us spoke French. We did not, but she quickly indicated a yacht heading towards that shore surf line; she thought it was drifting. Well, we knew it wasn't *Marmax*. As it turned out, the Ukrainian yacht captain was driving about in circles, waiting for his alcohol-seeking crew to return to the ship. It left us wondering, if the natives were nervous, how many boats had been lost in the past on this nasty shoreline?

The supermarket opened. Our two new Ukraine friends ran, I mean RAN, into the shop. I told them to run as we were also chasing alcohol in there. They raced around the tiny shop, panted up to the counter, both leaning forward, hands clasped…begging?! "Have you any whiskey?"

"No, sorry, no whiskey," said the shopkeeper lady, apologising.

"Have you any rum?" (Nearly on their knees…praying?!)

"No, sorry…no rum either. We have wine and beer." The shopkeeper smiled generously.

We felt for the guys; we could hear the disappointment in their voices and see the pain in their faces. They cleaned out the shelves of red and white cask wines, plus cartons of cold beer and took off back to their circling yacht. In the meantime, I had to ask the shopkeeper to go find some more alcohol out in her stores out the back. Lucky for both yachts, the *Aranui* supply ship had come in the day before!

It was time to leave these fair shores and head to the Tuamotus. It was 550 nautical miles (1017 kilometres) of perfect sunny weather and a gentle 14–25-knot east to southeast breeze with enough cloud around to stop us from expiring in the humidity. As we were now travelling with a mainsail permanently reefed down, we were not breaking records. Still, it was a pleasant run punctuated by full-time *Marmax* College, Les catching fish, taking watches, lifting weights, eating, reading and dissecting discussions of many books, and watching movies. I was on the second series of *The Handmaidens Tale*; kind of weird to watch this stuff in the middle of the Pacific. We would never have the luxury of time on land for this slovenly behaviour!

We were again having problems with the boat batteries not holding a charge, and the water-maker had ceased working, which was quite tragic at the time. Water rationing was very tight; not fun for us girls! Stopping off in Papeete, Tahiti, was an option if the situation continued. The thought of having to go to a big city marina wasn't exactly attractive to us…Time was running out to get *Marmax* home. We still had many islands, adventures, and miles to cover across reefs and oceans in the next eight weeks before reaching New Zealand. Fingers crossed; it would soon get sorted.

One morning, I thought, let's all pretend we have just started on an extended tropical holiday. It then won't seem so sad to be coming to the end of our fantastic adventure. We wanted to keep going; we couldn't. The cyclone season was about to begin for the Pacific. The trip back to the bottom of the world to New Zealand was a notoriously rough and often dangerous ride. We had no time to waste.

The good news was the batteries had been charged, the generator was working, and we were back into total water production on board. Pape'ete was off the list at this stage. We could press on to explore the Tuamotus, Mo'orea, Bora Bora, and possibly remote Palmerston Island if we had time. Niue and maybe the Kermadecs if the weather was decent, then home to New Zealand. Well, that was the plan, but plans can change!

Taioa Bay– Nuku Hiva

Fatu Hiva

Ua Poa

Glorious Hanavave Bay—Fatu Hiva

15. Tuamotus: The Pearls of the Pacific

Hands up, anyone who knows anything about the Tuamotus? Outside of our local yacht club, I knew very few people who had ever heard of them. It was also a bunch of islands that the *Marmax* crew knew very little of, too; who would imagine we would ever get there?

The other name for the Tuamotus is the Dangerous Archipelago. It was now quite clear to us why they called them that. Holey moly, we had some hair-raising, adrenaline-pumping passages through those atolls! The Pearls of the Pacific would be a much nicer name, don't you think?

There are seventy-six beautiful islands in the Tuamotus; they are said to be the largest group of atolls in the world. Lying in two, almost parallel lines, like two chains of silver necklaces. The place is stunning! Please do not ask us what country was the most beautiful now…Our *Marmax* cup had runneth over! The islands ranged from unbroken circles of surrounding coral lagoons to chains of coral islands with one or two navigable passes into the lagoons. The necklaces stretched from southeast to northwest for around 1000 nautical miles, 1850 kilometres for those not into nautical miles. There was little wonder why so many yachties had chosen to stick around for months and months. There was so much to explore!

The Tuamotus is where the French did their nuclear testing from 1966 to 1996. France moved its nuclear programme from Algeria to Mururoa Atoll and the southern Tuamotus. They then set off 193 atomic explosions in this pristine paradise. Grrrr! Thank God they came to their senses after the world rallied to condemn them. It is unbelievable anyone would want to wipe out such beauty. While we would think it would have been distressing to the inhabitants, the fact was that it apparently brought much-needed employment into the area for the locals. Before this employment, the natives had led a peaceful life on small farms. They grew their own food with their own hands, fished, raised pigs and chickens, eggs, etc. and happily traded. Here it comes again. A simple life of subsistence turned into a life of depending on money. The French nukers packed up their toys and vacated the area. There was no money left, and now, consequently, the French government is apparently still forced to subsidise this group's incomes.

The Tuamotus are the home of the Pa'umotu people who live with a coconut and fish economy. What did this mean for us yachties? Yep, as far as food was concerned, we had to be totally self-sufficient most of the time – there was only coconut and fish!!!

Though sad to leave the glorious heaven of the Marquesas, we had a great sail through to Kauehi, pronounced "cow-pee," our first atoll of the Tuamotus. We had read, in a cruising autobiography, that this place was well worth a visit. The entry to the atoll was exciting and, on first impressions, simply stunning!

This was our first taste of shooting ferocious tide rips in tight atoll entry passages. The water was see-through clear, the colour of Gatorade! You could see everything going on around the bottom. Sharks, manta rays, BIG CORAL BOMMIES, all racing under our keel as the ocean ripped into the little passages and the tide poured out from inside the atoll like a drain in a bathtub. We tried to shoot these gaps at slack tide with the sun well over our heads to spot uncharted coral heads. There were plenty of them here. We had heard a couple of horror stories about this pass, but it was no problem for *Marmax*; she is a powerful beast with a big propeller.

The coral reefs are dazzling from the ocean, but as you get closer, they become jagged, clutching death traps to any boat venturing too close. The Kauehi Gap was a total washing machine. Exhilarating, wild stuff. We popped out through the turbulence and into the lagoon. It was super serene inside…a beautiful big bowl of blue deliciousness, fringed by pure white sand and lush coconut palms leaning over it all. High fives all around, we had made it! Little did we know at the time, this was one of the EASY entrances we were to encounter.

We excitedly anchored up ten miles across the lagoon outside the village of Kauehi. Ha ha! Not much going on in this little ol' town – apart from us arriving!

"Where's the wi-fi?" moaned Janelle.

As usual, we had arrived on a Sunday, so the tiny town was quiet as a mouse. It always amazed us how quiet these little villages throughout the Marquesas and Tuamotus had been when we arrived. Lots of houses, very few people. One could not help but imagine they were all hugged up, backs to the wall, craning their necks from behind the curtained window frames, hands over their children's mouths, having a peek-a-boo at us. I bet a hundred dollars they were! We were a curious-looking mob at times. We could hear a couple of TVs going, snatches of laughter from a few guys on the beach. Found the local grocery shop with one to two cans of a few varieties of veggies, empty freezers, one bottle of Jack Daniels. Not much happening there either. The lovely lady had opened the shop, especially for us, so we bought three salty plums, a carton of pineapple juice, and a loaf of frozen bread for her efforts. *Marmax* was kept pretty well stocked, but we continuously foraged for fresh fruit, veggies, and meat. Oh, and alcohol, but in all fairness, we had only stored in the Marquesas five days before, and we don't drink that fast!

We were welcomed by colourful and fragrant flowers, a cute pink piglet, quirky-faced goats, roosters, coconut crabs, dogs, and a steaming hot concrete road during our wander through the village. It was a quick exploration. Clambering back to the boat, we took off to another remote part of the lagoon. These villagers may have been very shy and quiet, but they sure did live in a beautiful part of the world.

Most of the Tuamotus have a government-subsidised copra economy, and recently, many had gone into black pearl farming. The Kauehi people had many pearl cultivation nets lying about. There were fishing boats and coconuts everywhere, but the business enterprises looked like small family affairs by comparison to the following few islands we were to visit. They led a peaceful existence and were not geared up for anyone but the locals. It was lovely to see things in the raw. Unspoilt island living, and there is no way a cruise ship could squeeze through the plug hole into this place.

Whilst dodging fish traps and coral bommies, we zigzagged across the lagoon. Janelle had fun squealing all the way in the bosun's chair hanging off the spinnaker pole, getting regular dunks from Les. At the same time, Bruce was lurching the wheel from side to side, making *Marmax* rock 'n' roll. We hoped no one was watching our wild antics through binoculars!

We had arrived at the most perfect tropical dream island you could ever imagine. When was the director going to yell, "Cut!" Wow, what a treasure! Flat, coconut clad islands, fringed by white coral sand, protected by underwater jewels of colourful coral bommies up close to the beaches. The water was full of aquarium fish, all shapes and sizes, with blacktip sharks cruising as if they were guarding the treasure. It was nerve-racking trying to get used to swimming with sharks, particularly for Les and me. We had lived in Far North Queensland, Australia, for over thirty years with the man-eaters.

Another thing we were to find out: these sharks were tiny compared to the few islands we were about to discover. We snorkelled to our hearts' content; our skin was glowing like mahogany timber by now. The water was like champagne, too hard to resist any time of the day. We walked to the rugged seaside reef on the other side of the narrow islands.

Millions of years had left metres and metres of a deep, broken white coral on the coast; it was indeed a fantastic sight and so much fun to fossick along. The Tuamotus were pristine beyond belief. Apart from the rare plastic bottle, probably carelessly tossed off a boat, and the odd plastic fishing buoy, the miles and miles of coastlines we explored were immaculate. We often saw people raking and sweeping up leaves and pine needles on the other side of the road opposite their home along the beachfront. They then made a little fire and burnt their rakings. Proud people who pulled together to keep their village spotless.

At Kauehi, the beaches were a couple of metres deep with tiny, tiny white shells; very therapeutic to walk through. Now that I am back home in New Zealand, I keep a small bowl on a side table full of my special little Kauehi shells. They are kind of like a bowl of magic to me now, I guess. I run my fingers through those beautiful shells, and the memories flood back to that beach on which I found them. It makes every hard day seem so much better.

We reunited with an American couple whom we had met in the Marquesas, Ally and Patrick. We never realised that American jargon had so many differences until we spent several hours howling with laughter swapping nonsense. Ally had just found out what budgie smugglers were. We had no idea the Yanks called them banana hammocks. Did you? I also learnt the "lefty-loosey/righty tightly" rule for taps. Am I the only adult in the world who had not heard of this lingo? It is such handy terminology when running a farm with anonymous water taps all over the show.

Ally had also worked on cruise ships in Las Perlas, out of Panama. Remember the beach rubbish we reported there? Ally told us when she worked on cruise ships in the area, the crew used to go ashore at 3 am to clean the beaches before the tourists visited the next day. We were fascinated to hear this and rather pleased to discover that at least one part of the tourist community was cleaning the place up somehow.

A smart-looking guy magically turned up from the horizon alongside *Marmax* in a fancy sea kayak. Andy was in, at a guess, his late thirties and had come across to say gidday. He had ten of his family aboard his 46-metre (151-foot) superyacht, with twelve crew looking after them. Choice! They spent a few days anchored up at the Kauehi Pass, diving, spearfishing, and having a full-on masculine thriller holiday playing with sharks and shooting giant doggy tuna.

They must have been a fit lot to swim those hairy currents! Oh, the life!

Kauehi is part of the UNESCO biosphere reserves. It is one of seven atolls classified as reserves due to their unique and precious natural and cultural heritage. "Respect to protect" is the mantra.

As you can imagine, once we set anchor, it was always exciting to throw the tender over the side and get ashore to go exploring. Our next stop was to be Fakarava, another UNESCO reserve. After shooting the northern passage, (Yeeha!! Bouncy, bouncy!), we arrived at Rotoava, a pretty little village with the friendliest people on earth, always waving out from their houses and cars. "Bonjour" by the dozen down the main street; what wonderful kind souls they were.

This island clearly survived not so much on coconuts and fish but black pearls. We dodged several pearl farms on the way in, and ashore, small commercial pearl farms supplied tiny pearl shops and a couple of up-market pearl boutiques.

The colour of black pearls does not refer to the pearls themselves but the use of the black-lipped oyster as the surrogate incubator in what they call the Mikimoto culturing technique. Yes, I knew you wanted to hear about pearls! I'll shortcut my newfound knowledge! Pearl sellers there, by law, had to be government licensed, no matter what part of the process they were, be that a diver, jeweller, store owner, or street vendor. You had to legitimize your purchase of black pearls by buying from a reputable dealer. They issued a quality certificate, plus a bill of sale that had to get checked by customs as you departed the country.

Of course, we were confronted with street stalls at the markets that showed no sign of legitimacy. We tried to buy an exquisite pearl from one of the few dinky-di dealers, but I needed my passport to secure it. Unfortunately, my passport was back on the boat, so we had to return the next day. I returned five times back to that darn shop, but they were never open. Needless to say, they lost a good sale. I was momentarily satisfied by purchasing a few from the friendly street vendors. Certainly not the quality I was after, but every pearl has great beauty. The colours were a mix of what I would call Pacific greens, with a green/blue luminous glow.

We went into the post office to buy SIM cards so we could communicate with home.

"Sorry, the SIM cards are not working so well this week; I cannot sell them to you. Go to Fakarava Yacht Services." The big postie lady beamed.

No problem. I just did not like to hang around these establishments for the twelve to sixteen hours it took to whack out a blog! There was no choice here. Fakarava Yacht Services were marvellous hosts. All the visiting yachties hung out on the front veranda of their house, tuning into the world. At the same time, their family brought us cups of sensational machine-made coffee, Coke, and juices available at excellent prices. The family basically opened their house to everyone. The book swap was in their lounge! We swung them some thank-you beers later when we left the island. Champions!

The one and only ATM on the island would not take our cards, so Glenys did some bartering (begging) with a supermarket owner to get her cash. We met so many people. A couple of lively Danes on bikes on their annual holiday – "See you next year in New Zealand!" Lovely, lovely people who we bumped into a dozen times in the coming days.

We met Enaho, an artist with nine dogs and an art gallery on the beachfront. Apart from excellent eye candy for Glenys and me, the guy was a superb artist, and his waterfront house looked right over *Marmax* in the bay. There was a reason this guy came into our sphere of influence on this trip…more to come. We also had the pleasure of meeting Carola and Daniel, a German couple travelling worldwide with their nine-year-old son, Fritz, in their catamaran…more to come on them too.

That night anchored in Fakarava, during a howling gale, somehow, our precious inflatable tender decided to disappear. We had recently replaced her rope with a nylon floating line. Was she stolen? Surely not! No sign of it. Gone. She had untied herself and escaped. We thought there would be only one place she could be: up on the rocks on the far side of the lagoon. It was action stations. Janelle dropped Glenys ashore in her kayak so Glenys could hire a bike from yacht services to go search for this much-needed vessel!

"Go search the rocks outside the airport. We lose fifty dinghies a year here, and they all get smashed to pieces on the coral rocks outside the airport," the dive shop guy said. Great.

Janelle got back to *Marmax*. Les jumped into the kayak. He would search by sea, though it could have ended up miles away.

"Take your life jacket, Les, just in case," I suggested.

Remember, it was blowing 20+ knots, and the waves were up. We also had no communication with each other. Les took the portable VHF to call yacht services if the tender was found and/or salvageable. They were to bring a boat trailer to us if need be. Unfortunately, there were no VHF communications back to *Marmax* either. Our masthead aerial had previously gone for a flight to the moon with the Windex wind indicator back in Las Perlas Islands and still had not been replaced. Long story short. Glenys miraculously found the tender jammed under a coral reef ledge, being smashed to pieces. Lord only knows how she found it.

Les was not far off. When he saw Glenys (remember the wind and surf?), he searched for a spot to pull up the inflatable kayak. Coral jaws were everywhere. Side on to the shore, the waves tossed him out anyway. At least we now knew our fancy offshore category 1 survival lifejackets worked when they hit the water. I wish we had gotten a photo of him, but yes, the frigatebird look would be pretty accurate. As his life was being threatened, it was best not to whip out the camera at this stage! Sadly, the outboard was smashed to bits, but we did manage to salvage the tender. Yacht services hauled her up to the foreshore opposite *Marmax*, and Bruce patched her as best he could. She was now battered and bruised, and she deflated now and then, but she was to serve us well all the way home to New Zealand.

Enter the wonderful German couple, Carola and Daniel, off their catamaran, *Beluga*. Upon hearing about our plight, they took no hesitation in offering us the opportunity to buy one of their spare outboards they were carrying. They had three others: two biggies on the stern to propel their boat and one on their dinghy. Bruce and Glenys became the proud owners of a little 3 hp Yamaha. It did the job, albeit slightly slow, with five of us on board, but this was not a problem. We were so, so grateful to have had it offered to us. Outboard engines were like gold around there. Carola and Daniel would be in our lucky bubble again shortly; we just didn't know it at the time. Bless them!

The supermarkets were again, obviously, in French and incomprehensible to us. Glenys carried the translator app; I lacked the patience for it. I thought I had bought some good Angus steak. I swear that is what the label said. All meat purchased was always caked with grey-coloured ice, so that did not help. It was all imported New Zealand AFFCO meat up there, if you could find a supermarket open. We got back to the boat and discovered yet another dumb purchase of mine, sheep hearts! Despite my culinary skills in disguising weird food, the crew refused to eat them. Glenys came up with this bright idea: instead of feeding the marauding sharks, she would paddle ashore early in the morning and give it to Enaho, the sexy-looking artist with the nine dogs. Nice move, Glenys! Enaho was most grateful for our generosity. I don't expect the villagers would have been able to afford such luxuries for dogs. Actually, come to think of it, Enaho probably cooked himself up a feast that night!

Fakarava has two world-famous passes to access the atoll. Bold adventurers get carried in the tide, swept through the rushing water, and dive with the sharks, big mothers. Unfortunately, the weather was a little dangerous for us to do this. The "safety first" rule kicked in once the weather was doable. Family boat rules, we were pretty sure Mum wanted us all home in one piece!

Everybody was continually assuring us that the sharks were no problem. However, one of our new friends had a great story to tell us, and it was all about his experience in the south pass of Fakarava. We had met Sue and Laurie, two delightfully, adventurous retired Aussies from Fremantle, aboard their beautiful Catamaran, *Le Mitsu*. They had been cruising around for the past year throughout French Polynesia. Laurie took an American pal for a fishing session on the Fakarava south pass and got a shark on the line. You know where this is going, don't you…Shark got angry and bit a big shark-mouth-size chunk out of Laurie's inflatable tender. No thanks. It would have been a fair hike to get his inflatable back to safety from the pass with one hull totally deflated and two guys plus fishing gear on board! With a beaming grin on his face, he proudly showed us the panicky video clip he had managed to take of this fight with his angry shark; at least he lived to tell the tale. Good on you, Laurie!

The week before we got to Fakarava, a nine-month-old French yacht tragically got a dive buoy caught around its propeller going through the pass. Such bad luck. They were thrashed and lost the boat. They could not run their motor, and it was impossible to sail out of. Those jagged, clutching death traps had snared her. It could happen to anyone. Coral passes were a ton of fun when things were going right. Still, the word "fun" was really only mentioned once we had safely popped out the plughole into the paradise within the atoll. We passed another huge yacht high and dry in the coconut palms at the northern pass of Rangiroa; she was never going anywhere again.

We were due to leave Fakarava on a Sunday, but the supply ship, the *Aranui*, came steaming into the harbour. If you remember, we had last seen her in Fatu Hiva, the Marquesas. Our Kiwi contacts had mentioned our local Sandspit Yacht Club pals, Laurie and Jenny Barber, may be cruising in the area aboard this particular ship. Hold on a minute, what if they were on board the *Aranui*? It would be a shame not to at least check the possibility out! So, much to Bruce's teeth-clenching frustration, we delayed departure and dropped the tender down again. Glenys, Janelle, Les, and I headed ashore, with our Sandspit Yacht Club burgee (club flag) tightly rolled up in my bag. We were prepared for the goofy "Woo hoo! Welcome to Fakarava" thing.

As it turned out, we waited and waited, scanned the faces of everyone coming ashore from the supply ship and met some beautiful people off the boat. We chatted to the *Aranui* crew and handed our boat card to them if they happened to find Laurie and Jenny aboard. Who knows, maybe they were aboard another supply ship, in another land? We enjoyed many interesting speed-date conversations on the main street, all the way to the *Aranui* dock, where it was off-loading. The other big plus was all the locals had come out to play! The port suddenly blossomed with fabulous, traditional art markets. Janelle and Les found a family who made decadent waffles filled with Chantilly cream and chocolate. Oo la la! We had promised Bruce we would be back to *Marmax* by 10 am. Phew! On the dot. We lifted anchor, and we were off sailing to another atoll.

We enjoyed a perfect, peaceful sail to Toau and an easy passage into the tiny anchorage of Amyot. This was a blind passage, meaning you enter the passage into a little bay but cannot go further because it is sealed off by the coral reefs inside. There were already a few other yachts and catamarans anchored here. Ashore, the local hosts were a couple named Valentine and Gaston. They lived in an old fisherman's hut, right on the water. This couple were famous for their delicious fresh seafood dinners prepared for the yachties. Crayfish, parrotfish, raw fish, salads, and great big smiles. They owned two gorgeous dogs who escorted us out to the reef so we could explore. The dogs chased blacktip sharks, parrotfish, and hermit crabs across the shallow water and corals. They sure had tough paws. There were three kittens, a giant pet frigatebird, and ten penned pigs out the back. We would not have minded being around on pork crackle night at Valentine and Gaston's!

The snorkelling was absolutely phenomenal. Loads of huge, black coral trout roamed around the bottom with dozens of varieties of colourful fish, including weird unicornfish and ribbonfish. While on anchor, *Marmax*'s hull was being cleaned by around twenty suckerfish, accompanied by pretty dinner-plate-sized yellow and grey fish. We were not sure what they were called. It was all going on underwater here! It was still fun tapping Les on the shoulder while snorkelling miles from the boat and showing him the sharks on the bottom of the seafloor! Les is Australian, you see. Australian sharks are known as extreme man-eaters.

A German boat beside us traded rum and cigarettes ashore for a handful of black pearls. Our cruising guide we had aboard stated, "Don't think, just because you have consummated a smooth exchange behind a palm tree that it will not become known to the government. Informants exist on every atoll to relay shady dealings." What the? Quite unimaginable, but it just goes to show how serious they were about black-market pearls!

After a couple of nights in this stunning marine utopia, we headed off to the atoll of Apataki through the treacherous Pakaka Pass. WOW! That was scary stuff! A full-on, wild ride through. "How deep is it? How deep is it?" The rocks and coral under the boat looked so close, but it was twenty-five metres deep! The village of Niutahi was to starboard coming through – what a picturesque town this was. A glorious, clear blue lagoon that the entire village wrapped its arms around. We finally got spat out from the washing machine passage, totally freaked out by coral bommies and pearl buoys everywhere; did they drop down, spread across? Were they rope, were they wire? The sea was choppy with 15+-knot winds as we set anchor with *Marmax*'s stern to angry-looking coral.

We managed to surf ashore in the tender and took off for a jaunt around town. As was often the case, every business was closed. However, we did find an old lady baking baguettes in a half forty-four-gallon drum out the back of her stall. She had captured our attention because she was whacking the heck out of her loaves with a tea towel to cool them down. The smell of baking bread was glorious. A lucky find! We made her day by purchasing ten loaves from her for US$10 and were on our way with a cheery wave. We chatted to a few locals, so friendly, especially when they found out we were from New Zealand. Gosh, the NZ All Black rugby team were gods in these islands. Once again, Les and Janelle were devastated to go without ice creams again.

We just made it back to *Marmax* in the tender with our little 3 hp Yamaha. Straining with the incoming tide, we ripped through the fish traps, which we had to negotiate to return. That night, we slept with one eye open in the stiff breeze, stern still facing that nasty coral. It was too late in the day to navigate the uncharted lagoon to get to shelter.

This vast lagoon was loaded with pearl farms. At least two sets of eyes had to be constantly glued to the water surface to enable us to safely reach the next island inside the lagoon, Totoro. Pearl buoys and coral bommies everywhere! We wanted to check out Apataki Carénage Services, where there was a safe haul-out area for boats. In our travels, we had met a few Kiwis who left their boats here for nine months of the year and went cruising for the remaining three months. What a good idea; it had my brain ticking loudly. There were very few cyclones in the Marquesas and Tuamotus, making this place totally attractive for boat storage. The coconut trees and surrounding foliage were in perfect condition, a dead giveaway for proving it enjoyed good weather.

We had heard about a yacht smashing into Apataki reef recently. Here she was, safely on the hardstand. The poor guy! Apparently, the owner was teaching a few young men to sail on the outside of this atoll. He went down for a short nap (probably had been on a long haul), and after some time, CRASH! They had sailed straight onto the reef. The young guys had put the yacht on autopilot and were engrossed in a game of chess. Pretty messy; our hearts hurt for him. The boat was wrecked. Another badly damaged Bénéteau yacht was alongside. Another victim of that jagged and clutching death trap coral. The boatyard owner had bought the damaged yacht for US$1 from the insurers, who had written her off. Another sailing tragedy that had suffered from the loss of a watchman's attention on the wheel. Still, she was a lovely yacht, broken, bruised, and forlorn. If only someone could just wrap a giant fibreglass bandage around her hull, mend her like a doll. You could look through the jagged structure from one side to the other. These two damaged beauties were a stark reminder to us of the treachery of coral reefs and the damage they could cause. It was pretty confronting to imagine the grief the owners must have felt to have their hopes and dreams dashed with such a violent end. They say that sometimes the most important life lessons are the ones we learn the hard way.

There were many friendly nurse sharks in this bay because the family ashore had fed them their food scraps over the years. The giant beasts nuzzled gently along the edge of the sand, their huge powerful tails swaying with the incoming tide and baby waves. We went swimming anyway. Even sharks would not stop us from leaping like kids into the clear blue water, which bubbled like champagne around our bodies. We would be back in New Zealand soon, and this seemingly surreal paradise would be all but an unbelievable memory!

Apataki clearly had a much more sophisticated and lucrative pearl industry going on here. A Chinese company owned at least one of the massive pearl processing factories here. The pearl boats and crew were like clockwork with their work hours. This was quite unlike the islands we had previously visited, which seemed to take pearling as a casual pastime. There was an unusual-looking factory at the main village of Niutahi. I called it the Pearl Palace. I don't know what it was called, but it looked like a palace, a big one. Actually, if it had a couple of paddle wheels attached to its sides, it would have looked like a paddleboat on the Mississippi. It sat out on the water, away from the main island, we guessed as a security precaution. All staff and pearls came and left via boat. Quite a curious, out-of-place structure for these islands.

After playing dodgems with those pearl buoys, zigzagging across the lagoon, we found another gorgeous island beach to explore. We spent one more night there and then waited for the slack tide to shoot through the northern passage to exit the atoll. We were now heading for another atoll, Rangiroa.

Two hours out of Rangiroa, a little black cloud floated over us in the form of a vibrating/rumbling propeller. *Marmax*'s shaft had a knife-cutting contraption that sat in front of the propeller, designed to mangle and chop ropes before they went around it. We thought that perhaps we had run over some fish ropes or nets during the night. Anyway, the good news was after a few forward and reverses with the main engine, it sorted itself out. It was far too rough to dive on the hull in the ocean side of the atoll; the swells and waves would have knocked us out on the thrashing hull. That was one major crisis averted. Phew! More drama in the form of a stuffed fuel pump to run the generator was yet to come. No generator meant no water could be made on the boat.

We entered Rangiroa through the Tiputa Pass. What a buzz! We surfed through that passage on a low tide at 8.30 am. It was so nerve-racking to see cars, bikes, and people pulling up on the roads to watch us fly through. The dive boat operators were charging past us in the opposite direction, heading out with full boats of divers aboard, all tourists. It was incredible to see a couple of lads surfing along the coral ledge on our way out through the second pass when we later left Rangiroa. Wild! It must have been great entertainment for onlookers.

Marmax careered through that passage with much violent, smashing side-to-side movement. Once again, we were spat out into another gin-blue atoll with crystal clear water. Another lovely anchorage off a small village flanked by those Tahitian bungalows with thatched roofs that looked like they were floating on water, just like a postcard. Wow!

Let's go back to the subject of the generator. We were lucky to have two knowledgeable men in the bay, one a charter boat captain, and again, German Daniel off *Beluga* to the rescue. Enter the wonderful world of travelling yachties, the pay-it-forward system whereby everyone helps everyone. So Bruce and the boys came up with a solution to bypass the now defunct fuel pump. We now had a gravity-fed plastic tank of diesel sitting under the saloon steps and no lid on the bottom galley steps, as this was the hole that the jury-rig fuel lines came from. Glenys thought it would be wise to place our bag of seven long French baguettes on the floor next to the step hole so that we did not accidentally step into it during the night. Hmmm…how did that work? (It's 2 am, and I'm coming off watch. Now, where do I step? In the step hole or on the bread?) Got to say, the girl is practical! It worked; none of us fell down that hole.

We had a very long, hot, and fruitless walk looking for a petrol station to sell us a five-litre plastic drum of diesel for the jury-rig experiment happening on the generator. No joy. Why did we always end up at airports in the middle of nowhere with nothing ever open? We did find an ATM that worked – TICK! Trudging wearily back to the tender, Glenys and Janelle scored a ride with two lovely ladies then came around the corner and picked Bruce and me up. "Janelle…did you do the limping foot while leaning on Mum trick?" Next time we make an ocean voyage, we are taking fold-up bikes!

We took a taxi boat to explore another island. There we found some supermarkets that were actually open and well stocked. Later on, to cool off, we swam with the delightful local kids at the main dock. Back at the boats, we enjoyed the company of newfound friends Laurie and Sue, the Aussies with the excellent shark story. Swapped stories, pictures, secret wi-fi codes, USB drives, yacht information, and, of course, sundowners on the deck of their beautiful catamaran. The men spent hours on the *Marmax* generator! We were to head to Makatea, but all good sailing plans do change for various reasons. We now had the generator working without the fuel pump, and on a gravity feed, so we set off for Mo'orea. Now that was a place we had always wanted to go to! Out biggest fear? We were heading into civilisation again.

Sailing past the last Tuamotu atoll, we reminisced on the past couple of weeks of exploring. The Motus had been mighty impressive. Untouched and unspoilt. The vastness of the lagoons had been astonishing, many with no horizons. The simple lives of the natives reflected their lack of need for anything; they were genuinely content. Always smiling, always gentle and helpful. They watched out for us and our boats and never pestered us.

The reefs on the outside of the atolls were ferocious, with seas crashing over big walls of iron-hard coral. The inside of the atolls, embellished with coconut palms, had always rewarded us with brilliant, white beaches and pristine waters going from emerald green to a deep blue, dotted with those luminous sand cays, pristine coral, and fish life. We were often left speechless.

Best of all? We had been somewhere where most people do not go...because they can't. Cruise ships could not get through these tiny passages into the vast lagoons. Only the lucky few in private boats with a strong sense of adventure were the privileged.

8 pm 19th August 2019. Mixed emotions – the city lights of Pape'ete were glowing on our port side..."Ground control to Major Tom."

Kauehi Atoll

Fakarava

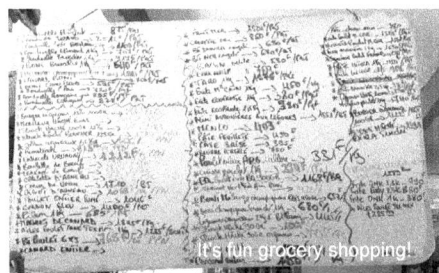
It's fun grocery shopping!

Peaceful, remote Kauehi

Having a bit of fun zig-zagging across atolls

We find our very smashed up outboard!

16. The Society Islands: Stuff Dreams Are Made Of

Time was now whizzing by for us as we sailed down the South Seas, getting closer to our home of New Zealand. We had left the stunning Tuamotus and sailed onwards to the Society Islands. Mo'orea, Tahiti's little sister, is a glorious-looking island and obviously very famous for it! We felt like we had been sailing about inside a postcard; it was totally surreal. Travelling towards Mo'orea was so much like sailing up to the magnificent islands of the Marquesas; even in the semi-darkness of early morning, the mysterious, jagged mountains were breathtaking. Because of our early morning arrival in Mo'orea, we had to sail about for several hours outside the lagoon to wait for the sun to rise to enable us to safely navigate. Gliding through the reefs into the lagoon of Vai'are, we were thrilled to be welcomed by spinner dolphins doing their thing as we passed through the port and starboard passage markers.

We dropped the anchor just short of the inner coral reef outside the little township. This was to be the first large, commercial supermarket we had found for many, many months, actually since Gibraltar! Yes, we were like kids in a candy shop; they even sold huge flat-screen TVs. The same cut, first-class NZ AFFCO meat was the same price, or cheaper, than what we got locally in Auckland; why was this so? Anyway, that place was fun, and rum, at long last, was coming down in price; now US$10 less a bottle than the Tuamotus.

After navigating our way through more sharp coral to get the tender ashore, we observed not only seaweed on the beach but brown/black volcanic sand. The Tuamotus were so different to this. We walked down the road for a wander around the Vai'are Marina, Mo'orea's only marina. What a sad sight that was, even though it enjoyed the most stunning picturesque backdrop. Its eighty-five-berth capacity was like a graveyard full of yachts. At least half of the vessels were in varying degrees of decay and suffered a great lack of tender loving care. Mouldy old sail covers, spray dodgers, and biminis falling to bits. Shredded sails flailing on the masts. One mast had been completely snapped off at the top a long time ago. The DIY addition of a sugar scoop (stern of the boat) was now falling off, teak decks curling their toes up, rotten ropes and anchors, so deteriorated and corroded. There were many live-a-boards; I guess this is where the term "grotty yachties" originated. Nothing like our sharp and shiny floating beauties back home, quite an eye-opener for us.

Would we park a yacht in here? Err…probably not. We would be concerned there would be nothing left of our shiny bits on deck if we went away for a day. We made a mental note to lock *Marmax* up very tight here. In the days we were in Mo'orea, we heard of a couple of visiting yachts that had been broken into. Some of these swarthy, dirty-looking, barefooted boat owners looked a little suspect. So as it turned out, our paranoia was justified; in saying that, we had never left *Marmax* unlocked when not aboard. The marina itself was very smart, had good security, crystal clear water. How would they get rid of floating vagrants or rotting carcasses of boats when they had to? I would have liked to have chatted to this marina manager, but it was time to move on.

Out on the road, we fell in love with the rest of Mo'orea. Friendly villagers had road stalls lining the main street, selling their home produce of coconuts, glossy avocadoes, cucumbers, melons, pumpkins, and sweet-scented pineapples. One stall was laden with stunning, intricately weaved flower leis and heis (head leis). The fragrant perfume of that stall was divine! Frangipani, bougainvillaea, wild orchid, and tiare blossoms, which are part of the gardenia family.

I had been trying to transfer money to a mortgage account in Australia from New Zealand. Do you have any idea how difficult this task is while on a sailing adventure? The banks suddenly decide a security breach is underway because your computer is beaming out from French Polynesia or some strange foreign country. All security messages and PINs, etc., come through your phone. My phone SIM card had refused to work since leaving Galápagos. Les promised me that French Polynesia would have my carrier, Vodafone. I had severe doubts about that statement from him.

So here I was, gazing out across the harbour, and there flies a giant red and white passenger ferry with signwriting in bold lettering, fifty feet long. What do you think it said? VODAFONE! There were a few shops in Vai'are, but of course, one was red and white, yes, a Vodafone shop! Sorry I doubted you, Les. You were right. Vodafone? What Vodafone? They could not fix my SIM card problem, so Les and I walked away US$180 later with a new card full of data credit. At least communications were up and operating for me now.

Glenys popped her head into Avis car rentals. They only had tiny 4 person cars, and there were five of us. She found a lovely big, ukulele-strumming man in a colourful pareo behind the Avis desk. "We would like to take a tour around the island. Do you know of anyone?"

The big dude phoned Mister Joe, who, in turn, called Delores. Delores, our chatty driver, turned up thirty minutes later in a nice, big, air-conditioned people-mover. Perfection! For US$250, Delores cheerfully gave us an animated, incredibly informative, and hilarious round trip of Mo'orea. We knew she would be fun when she mentioned in her briefing that we would be visiting the Fruit Juice Factory in a village called Piha'ena. She gleefully told us this would make us very happy for the rest of the tour.

"Come on, Delores; let's do the Fruit Factory thing first!" we sang out. The Fruit Factory was a front for extreme alcohol tasting activities, in which we partook with gusto and enthusiasm. We unloaded a significant amount of money into bottles of the stuff when we left to keep that happy feeling for the rest of the sail home.

Delores took us up 'Ōpūnoho Valley, the centre of the volcano that had initially given birth to the island of Mo'orea. Through a winding road, lush rainforests, ancient maraes, paddocks of well-bred cattle, vertical rock faces, and incredible mountains, up to the Belvedere viewpoint at Mount Rotui, the sacred mountain of the ancient Polynesians. Spectacular bays and valleys spread out from the top: pineapple, vanilla, coconut, and banana plantations, rivers and cascading waterfalls. The surrounding forests were just pristine. This place is out of this world! Delores finished our tour off by showing us where the blue-eyed eels lived under a bridge. Huge big things. Thanks for the beautiful day, Delores!

The next day, we rode the big Vodafone ferry across to Pape'ete, the economic heart of Tahiti. At 34 cruising knots, this trip across the water only took twenty minutes. Racing along at such speeds across a major highway for whales and canoes, I guess this was the smart thing to do: scan the water for obstructions! It was interesting to observe the spotter crew on the top decks.

We could not come to Mo'orea without visiting Pape'ete. Expecting high-rise buildings and millions of people, we were very pleasantly surprised. The city buildings seemed to have a height limit of only four storeys, and only around two hundred thousand people lived there. The highlight for us was the Pape'ete Market. I loved this place and would have been happy to fill a forty-foot shipping container and send it home, but alas. I guess it's a girl thing. I found Les having a snooze on the lawns outside a church in the middle of town, attentively guarded by a dozen or so scraggly city chickens.

The markets covered seven thousand square metres, displaying a mind-boggling range of authentic, tribal Polynesian art and craft. Fruit, vegetables, amazing flower bouquets, handicrafts from all the islands, woven hats, bags and baskets, black market pearls, and shells. Marquesan stone and wood carvings, tifaifai (patchwork), jewellery, decadent French pastries, you name it. The markets were so colourful, so tasty and crackling with energy. It was interesting to note the high-security measures they discreetly carried out. I suspect they were also not without their social problems. Undercover detectives, eye-in-the-sky cameras, a lady disguised in gym gear mingling with a young crowd with secret-service-like earbuds connected to a walkie-talkie under her towel. There were heaps of them. Pape'ete was indeed a culture shock to our quiet, oceanic dream lifestyle!

Traffic jams, modern clothes, women wearing make-up (we had not seen this for a long while!), men wearing make-up. One of my best sights was seeing eight fa'afafines, or fafas, the boys raised to be girls, in full regalia. Beautifully dressed and coiffured, moving those slim hips in a wicked swagger like they were strutting a red carpet on a world stage. Female overload! I must say, it was rather enlightening, as a woman, to sit in a cockpit with other ladies, always barefaced, au naturel. No nail polish, interesting grey and blonde bits coming out through our hair, our clothes of creatively mixed, bright colours and varying stages of disrepair or, as aboard *Marmax*, nicely hand-sewn back together. Stick a flower in your hair, and baby, we all looked great! I personally could have easily gotten used to this.

Interestingly, all marketing centred around the images of beautiful long-haired girls with flower leis around their heads and gorgeous explosions of colour in their pareos. Yes, that is just what they looked like too. Church day was magnificent, with all the ladies dressed in their finest. The stallholders, the girls behind service desks, even young girls swimming; they really embraced their femininity there. I could not help but notice that it was all geared up for women. I mean, men don't wear black pearls either. Even the local Hinano beer had a young girl as their famous logo, and it was fascinating to learn she had also been adapted to modern marketing tastes over the years with a morph from a young, hunched over, slim, dark girl to a slightly plumper, definitely older, and paler girl.

I was one lucky girl in Pape'ete. Les bought me two stunning class A pearls; I was thrilled to bits! Les's birthday was coming up, so we hit the streets to purchase his Tahitian resort wear, not to confuse it with his British Virgin Island fashion! I always say it: the man should have been a girl.

The murals on some of the buildings were creative and thought-provoking, some of them just beautiful to gaze upon. We took the last ferry back to Mo'orea that night. We made it back to *Marmax* just as yet another glorious Tahitian sun was setting. The whole horizon just glows brilliant orange and, in the background, the distant twinkling lights of Tahiti. Just magic!

The following day, we were so excited to be moving around to the famous Cook's Bay. This truly was the stuff dreams are made of. Cook's Bay was framed by those well-photographed, jagged mountains and had been famously painted throughout history by artists worldwide. Captain Cook anchored there, way back in 1777. The only problem with Cook's Bay was dogs barking all night. Bark! Bark! Bark! The inevitable sound of roosters early in the mornings seemed to have followed us since the Mediterranean; there were now just more of them. The place sure was a nice spot to take memory photos of *Marmax*; I shall treasure them forever.

We filled *Marmax* with fuel at Mobil Paopao, the prettiest fuel dock you will ever see in your life. The second morning in Cook's Bay, we moved to another spot in the bay, which for some reason, seemed very popular to the other yachties. Was it the fact the dogs could not be heard here? Or was it the great wi-fi in the tiny garden café up the road? I'm sure the topless Tahitian lady ashore was not a diversion. Or the joy of finding an extremely rare, freshwater, outdoor shower ashore for those boats without water-makers. Everyone seemed to be having problems with their water-makers but, for a change, not us! If we gave our jury-rigged generator a few heart pumps of fuel now and then, it would keep its gravity-fed pressure up. Consequently, the water-maker was now alive and well, unlike the inflatable, which needed regular heart pumps to keep her sides up!

We awoke one morning surrounded by a curious, thick seaweed. Now we have not mentioned the state of the coral reefs in Tahiti yet. Basically, because they were, so far, totally unspectacular. More on this later. They had this weird pine-cone-like seaweed attached to a lot of the coral here, plus a softer seaweed that had millions of fertiliser-size seaweed balls attached. Add this to all the tiny nuts and coconuts, and you end up with a giant floating seaweed mat caught in the tide, full of all sorts of exciting things. Swimming amongst it, I must admit that I was a little worried that it had formed some kind of fish attraction device. What would lurk under this stuff? Why would you swim with it? No need to worry, the water was so, so clear. The boats looked like they were suspended in air, a little like plonking your boat on cling-wrap over a bowl of water. The clarity was exceptional!

We had Les's birthday celebrations with a chocolate cake made up with ingredients from four different countries. We then had fun re-discovering and unwrapping all his clothes from Pape'ete and spoiling him with more chocolate. A great night was had, catching up with our Kiwi mates aboard *Allesea*, whom we had first met in the Marquesas; remember the Kiwi rescuers?

Unfortunately, because we were speed-dating half the world, we eventually had to leave this glorious morsel of Mo'orea. A few humpback whales put on a fabulous display, and Cook's Bay farewelled us with a lovely little rainbow. Yes, we were feeling the love from Mo'orea! That was some special place, and yes, right up there with the most beautiful. But the trip was not yet over!

A sloppy, windless overnight sail to, oh boy, yes, another hand-on-your-heart beauty, the island of Huahine. Another exciting welcome for *Marmax*! Orca whales glided by. Humpback whales, 1, 2, 3, 4. What was going on? Sails down, we edged over to the action. There was clearly a scrap going on here. Were the humpbacks nursing calves? Were the orcas circling and hunting after them? Maybe the whales were giving birth, and the orcas were eating them…next subject.

Hard to imagine but, to get there, these fantastic mammals had travelled around seven thousand kilometres from Antarctica. They fed on shrimps and small fish during the summer, down there, then came up to reproduce and give birth in the warm French Polynesian waters from August to November. On average, they were a whacking fifteen metres long and weighed in at an average of twenty-five tonnes. They can live longer than forty years, are pregnant for eleven months, and give birth to a whale calf that will pop out four metres long at about eight hundred kilos in weight. Apparently, those spectacular jumps out of the water are loving parades performed by the females. While the girls are doing their thing, the male whales sing to them. There's your short humpback whale lesson of the day.

The morning sun dazzled over the mint green surf, smashing upon the coral reefs on both sides of the lagoon. We thought we'd experienced super-clear water before, but Huahine was to trump them all. Huahine was 170 kilometres northwest of Tahiti. The white coral sand on the bottom glowed up, and every fish, every piece of coral was on show. *Allesea* joined us in the bay; it was about to be a great night ashore to continue celebrating Les's birthday. Our anchor spot, Fare, was a small village that appeared to be in the throes of becoming a more sophisticated tourist spot like Mo'orea. They had a long way to go on that front, but they were still very much set in their authentic ways. The people, so friendly with magnetic personalities. There was lots of road development going on.

It was gratifying to study the rainforest of these islands. Huahine was a true Garden of Eden, from the tips of the rugged mountains all the way to the luminous edge of the clear water: thick, lush, pristine rainforest. After living many years in Northern Australia, we had become used to seeing the long-term, devastating effects of tropical cyclones on vegetation. It never really recovered back to a state of perfection like this. Never. If you close your eyes and imagine the mountains and valleys of a million years ago, I'd be willing to bet it hasn't changed at all in Huahine.

We found, what I initially thought, was a sex shop. Well, it was called Distillerie Huahine Passion. In my defence, the bush was covering the "Distillerie" word from where I was standing. Bruce's guess was correct – it was a distillery! No comment. Somehow, we were vacuumed into that little door, and a friendly young man plied us with shot glasses of a variety of alcoholic juices. It was safe to start with the easy stuff, then we started downing divine liqueurs and rums. One after the other, after the other…Alcohol percentages began at 18% and ended at 55%. Holey moley, do not light a flame in here, folks! It was of excellent quality, so we bought a couple of bottles, mindful we already had a box of super-charged booze from the Fruit Juice Factory of Mo'orea. Drinks for special times. This tasting experience was a good heart starter for the rest of the night with Kate and Ian, starting with piña coladas all around. Gosh, we love this life! It was a happy birthday to Les (again…hic!)

We motored on down the coast inside Huahine's magnificent reef system. The islands here were surrounded by patches of coral, with a centre of deep blue channels of water, then the outer reef. Stunning, addictive colours we will never forget. On the way, we passed one of my favourite mooring buoys, which I had fallen in love with within Panama, only this was a click-clack plastic set-up. Not quite as hunky as the big rubber ones we had previously tied up to. Moored in the deepest part of the channel, so obviously for large vessels. Later, we would find the mighty four-masted *Wind Spirit* tied up to it during our return to get back out of the lagoon. She looked like she could be a bit of fun! She was based in Tahiti, was 360 feet long, and carried 148 passengers and 101 crew. (That is a good ratio for service!) A seven-night cruise on this beauty at the time was only US$1967, with beverage packages at $32 to $52 a day.

Down to the ethereal Avea Bay for snorkelling. What a b-e-a-u-t-i-f-u-l place! The colour of the water was unbelievable. It seemed like we were suspended above it. Two little floating houses lived in the bay; a few visiting yachts were hanging about. Surreal. It was interesting to observe evidence of coral grafting going on here when snorkelling. We had also found the first "Nemo" fish of our trip and the weirdest underwater pink anemone that we had ever seen. Glenys and I went ashore and checked out the tiny little village, which was bursting with, what seemed, every tropical flower and fruit tree on planet Earth. There was that Garden of Eden thought coming on again. We were beginning to feel a little anxious about leaving this fantasy land. Farewell to gorgeous Huahine; you too were up there with the very best! As we left, yet another humpback whale dived behind our stern, another marvellous farewell to the lucky *Marmax* crew.

A smooth fifty-kilometre sail across to the island of Raiatea found us in a very unfamiliar environment. It started raining! What??? It is well known that Les had this immense passion for his fishing and clothes, but did you know about his obsession with wet-weather gear? Honestly, he would whip his rain jacket out the minute a raindrop threatened. He'd snuggle into it, take charge of the wheel, happy as a pig in mud with a giant grin on his face!

The entry to Raiatea, Irihu Pass, was between two motus (islets). You must see these islands of perfection to believe it. It was a bit of a shame that it was raining, but we could not complain after six months of being blessed with excellent weather. I would have loved to have taken sunny photos of them in their splendour. We anchored in Fa'aroa Bay at the mouth of the Fa'aroa River, the only navigable waterway in the Society Islands. This bay was reputed to be the departure point where the Maoris originally left for New Zealand via the Cook Islands. The history here was massive.

With a slight clearing of the skies, Glenys, Bruce, Janelle, and I jumped into our soggy tender and took off to explore the river; it looked pretty enticing from aboard *Marmax*. We were lucky enough to come across a friendly villager in an old kayak at the river entrance who spoke English. He called himself James. Where did he get his superb grasp of English?

"The river is my university. For three years now, I have been speaking to all the travelling boats. I now speak English, French, German, and some sort of Asian." He grinned.

Plus, his native tongue of Tahitian. We asked how far we could go up the river. He was a wealth of information and very passionate about his river.

By the way, "My brother, André has a beautiful garden up there. Go see him; he will show you around."

We no longer need to see the Amazon River. Apart from no monkeys or crocodiles, we were sure this river may have been the same, or perhaps better. Much to our amusement, we passed a dog happily chasing fish in the middle of the river, then motored up a kilometre or so of unfolding, astounding beauty. Loads of breadfruit, fragrant mangoes, bananas, and swaying coconuts. The yellow blossoms of the cottonwood flower trees formed colourful umbrellas over the cool waters. So calm and serene. It was no surprise to find a beaming André waving his welcome on the side of the river upstream. We tied up at his rickety little jetty then spent the morning with this warm, witty man who showed off his colossal fruit, vegetable, and ornamental gardens. The soil was so deep and richly volcanic. The only pests were wild chickens and these two-inch round snails sliding about. They looked like they would be more comfortable crawling around a rocky reef than in the middle of brown dirt; they loved hanging out in the taro department. André filled our arms with sweet sugarcane, crunchy snake beans, pawpaws, limes, cucumbers, and coconuts.

"Sorry, André. We can't bring bananas on board *Marmax*."

Probably thought we were crackpots, but anyway...

André intrigued us with his weird fruit, including the disgusting noni fruit that Glenys, a good sport, heartily bit into on a dare. I've tasted it before, so I knew it was gross. Still, its medicinal and revitalizing properties are renowned in the natural remedy world. We spent a hilarious hour trying to whack coconuts and limes out of trees. André fashioned a stunning native headdress for Janelle from leaves and tropical flowers. Curiously, he also pressed the back of a fern against Janelle's black shirt, leaving an imprint from the white natural powder on the underside of the frond. The result? A perfect impression of the NZ silver fern emblem! We had a lot of fun and paid him well as a token of our appreciation. We had been lured into the family business by the customer service department at the mouth of the river...Smart!

We met another two crews aboard German and Swedish yachts anchored alongside. They had travelled from their mother countries a couple of years back and were making their way to New Zealand for Christmas. Swapping wi-fi codes that belonged to restaurants and hotels was one of our pastimes out there. On this occasion, the Germans offered us the wi-fi code of a resort over in Taha'a; apparently, it had the strongest signal since the Bahamas. Hold that thought; it came in handy two days later. Gold!

Rain, rain, rain. Sadly we had to cancel a planned trucky tour around Raiatea. We headed up the coast and came upon three marinas chocka full of charter boats, hundreds of them! We were seeking shelter from building winds. Raiatea was a total contrast to the previous islands we had visited along the way. It had the second largest population in the Societies after Pape'ete and was the administrative capital of the Leeward Islands. So much more developed, lots of European-style homes adorning the waterfront. It reminded us of canal housing in Queensland, Australia. The inhabitants could just jump in their boats parked at their front doors. Many wealthy French people had holiday homes here. We felt like we had just opened the wrong page in a book at that place; where were we again? It was weird to be anchored in a bay surrounded by housing. A pleasant lifestyle for some, but a little too civilised for us intrepid adventurers, so next morning, we travelled inside the reef to the beautiful island of Taha'a.

We dropped the anchor to go check out the island. The water was extremely deep around there, but we had a lot of anchor chain. There was not a lot happening in Taha'a. A few skinny dogs, chickens, and lush plantations. We had an awkward hand-gesturing conversation with a family merrily laughing and chattering away in their native tongue while working on dehusking coconuts. This was the first island we had come across that was clearly going underwater. The houses were already in line with the water surface of the surrounding ocean; the inevitable was happening. It seemed so sad to think about how their lives would be in the future. We went for a wander, searching for someone to give us instructions on how to get to this famous diving spot we had heard of. Apparently, you could start at one end of a pass and snorkel the whole thing between two islands. It was rumoured, amongst the yachties, to be quite magnificent. We found our source of information while partaking in the only food for miles, ice cream. Rolled vanilla ice cream. Delicious!

We found that elusive snorkelling spot, and WOW! Was it worth it! The Coral Garden of Motu Tautau was indeed magnificent and, yes…better than Toau in the Tuamotus. Smaller fish but the colours and shapes of the coral, the friendly fish, colourful corals, and clarity of the water was unbelievable. Spectacular coloured moray eels hiding in crevices, crimson octopus and fish of all shapes and sizes. Truly an unforgettable experience. Much to Bruce's total disapproval and frustration, Glenys, Les, and I hassled him into allowing us an hour ashore to go snorkelling. The three of us braved the blustering wind to get the tender into the spot. Unfortunately, Janelle and Bruce missed out, as they did not want to come with us; they are like two peas in a pod. Mind you, I, too, would be nervous about leaving *Marmax* in that weather; someone had to stay on board. The wind steadily increased, and where we were anchored was way too dangerous to stop the night, with big coral bommies on the seafloor and *Marmax*'s stern pointing towards the reef. (Remember the jagged, clutching death trap stuff? Aha, that stuff.) Super spooky, as the surf spray spumes up high around here, making it look even more dangerous. Off to another sheltered bay for the night.

Bora Bora looked totally tantalising from here, as in the distance, we had her famous mountains on the horizon. That wi-fi code from the Germans? Awesome! The weather closed in. There was nothing to do but hitch onto that wi-fi and catch up with the world.

During sundowners, we were entertained that night by two young American couples on a charter catamaran who tried picking up a mooring alongside us. They took the buoy far too fast, lost their boat hook, one guy jumped over the side while the boat was in gear (yikes), and he could not swim particularly well. His other half waved a life ring over the side in case he needed it. Saved the boat hook, had another crack at it. Missed it. Third time lucky and a big clap from us.

"Where is the nearest pizza place?" they shouted over the wind.

Pizza place? HERE? Dogs, roosters, but not pizzas. Poor things, but off they went, off into the sunset. We'd love to know what they found for dinner!

We left the beautiful island of Taha'a, motor-sailing on a blustery day, with the mainsail up to steady the rock and rolling of the boat. It made for another exciting passage through the reef passage, though. Once we were clear of danger and had set our course, it was a superb reach towards the island of Bora Bora. The magnificent 65-metre, three-masted British-registered schooner *Adix* sailed past in her splendour. This was the vessel previously owned by infamous Aussie Alan Bond, who had previously named her *XXXX* after his beer. Fourteen crew, ten guests, she was on charter. There was a little too much wind about to comfortably sail with a complete set of sails this day, but we imagined she would be quite a sight. Even at anchor, she was a true beauty.

By now, we had all donned long-sleeved T-shirts; it was already getting cooler. We were down to 24 degrees and now heading into Bora Bora with…gulp, only twenty-seven days to go before we needed to be home in New Zealand!

Cooks Bay—Mo'orea

Pape'ete Market

Les turns 68 in Mo'orea

17. Bora Bora, Palmerston, and the Paradise of Niue!

I lazily lolled about in the sunny cockpit of *Marmax* after my fourth swim of the day. The waters were so clear; it was beyond belief. The surf was gently rolling over the coral reef alongside us; the water was shades of scintillating blue from light turquoise through to a deep sapphire. Waving coconut palms on yet another stunning white sandy beach. Like they say, "the quintessential tropical dream." Oh yes…we will miss this! We were all now lamenting how much we were going to miss this incredible sailing adventure.

Bora Bora is only 150 nm (277 km) from Tahiti and is the oldest Society Island, at some seven million years of age. She is at the age stage of being halfway between the atolls of the ten- to forty-million-year-old Tuamotus and the two- to three-million-year-old high-peaked islands of Tahiti and Raiatea.

We surfed through the one opening of the Bora Bora lagoon, Teavanui Pass, between two beautiful motus. The tide fair ripped out that pass! The famous high-peaked mountains of Mount Otemanu and Mount Pahia rose up majestically before us. Even on a pretty ordinary, drizzly day, it was quite a thrill to finally gaze up at those mighty peaks from our yacht. We had made it!

The main island of Bora Bora is only 8.85 kilometres long and 4 kilometres wide. It can be driven around in no time. The road is only twenty kilometres long. This surprised us, as we had imagined it to be so much bigger! Unfortunately, you cannot circumnavigate a deep draft yacht around the island, as the lagoons are shallow in parts. The main island of Bora Bora sits in the middle of a massive blue lagoon surrounded by stunning coconut-clad motus. This is inside a magnificent barrier reef on which the ocean swells smash into plumes of white spray. Bora Bora's lagoon has a reputation as the most beautiful in the world. We needed to find out what all the fuss was about, so we jumped on board the free airport ferry (nice find, Glenys!), and we went for a ride from the main village of Vaitape to the airport. The Americans built the Bora Bora Airport on its own little motu during World War II, 1942–1943. The Japanese bombed the American Naval Base in Pearl Harbor, so a new base was set up here at Motu Mute.

Travelling by the airport boat shuttle was a great way to check out a large part of the lagoon and step momentarily into the shoes of a first-time visitor to the island. Talk about "You had me at hello!" We surmised the average Joe fell in love with Bora Bora the moment they got off their plane. Entering the airport, eager visitors are greeted with gorgeous, fragrant fresh flower leis and a bottle of cold water by their respective smiling hotel hostesses. Just a few steps forward, they are looking point-blank into the brilliant turquoise of that stunning lagoon. If you were lucky enough to be staying in a flash establishment, your hotel would pick you up in a classic motor launch. Very stylish, very 1950s. At an average tariff of US$980 per night, per room, I guess you would not expect to swim! We were advised that the average hotel room on the island was a million Pacific francs a night.

The looks of excitement on the faces of people exiting the arrivals hall were beautiful to watch. No stressed-out businesspeople galloping out of here! This was a one-coffee-and-a-cake stop for us, as we had to return to Vaitape on the return ferry. It sure was fun. The ferry flew past *Marmax* at anchor, all alone under those big mountains. I pondered from the top deck of the ferry how insignificantly tiny *Marmax* seemed. To think she had sailed us all this way around the planet; no wonder some of our friends thought we were crazy! It pays not to overthink these things.

The next day, Glenys found a lovely lady with a big taxi van. She was willing to take us for a ride around the entire island, yes, all twenty kilometres of it. By this stage, we had found Kiwis Kate and Ian off *Allesea*, so they joined us with their daughter, Millie, for a jaunt around the coastal road. The weather was a little miserable, and our taxi lady solemnly informed us that it was because the "rain boat" was in the harbour. Apparently, whenever the cruise ship the *Paul Gauguin* pulled into the main harbour, it always hosed down with rain until the boat left. What a shame that she was anchored alongside us for two days!

Honestly, we weren't as excited by Bora Bora as we were hoping. We were a little disappointed by the amount of rubbish on the island, considering tourism is the hand that feeds them. It does not take much effort to keep streets, bins, and watercourses clean. More action is required, Bora Bora! I understand such a comment is controversial, but seriously? We had travelled halfway around the world to find, of all places, Bora Bora in a sorry state. The locals of Mo'orea had jokingly called Bora Bora "Boring Boring". We probably would not go that far.

Known as "the most beautiful island on earth." Hmm…good marketing, we say! Not sure we would agree with that statement at all.

Three major hotels on the island had gone bankrupt in recent years. Though set on idyllic, beachfront real estate, these places had gone to wrack and ruin, chained up and left to, disappointingly, rot. Morbidly fascinating but terribly sad to see. Bora Bora's main town of Vaitape boasted a great docking area for vessels. The main artisan markets were an absolute delight, though often triple the prices we had found on other islands. Pearl shops and art galleries were sprinkled up and down the central part of the village. You would not think you would have to navigate through traffic, huge potholes, and muddy roadsides to get to them.

Out of the town, however, the swimming beaches were magnificent. Silky white sands and water, still that jaw-dropping crystal-clear blue, the central mountains' twin peaks always dominate the background.

While wandering about the village, Les and I discovered a talented French painter named Jean Paul Frey. His brilliant pieces depicted scenes around Bora Bora and collages capturing the tropics' vivid tropical colours and romance. We both gave our bank cards a bit of a thumping in there. If we didn't, we knew we would walk away and forever lament over that decision.

One cannot possibly sail to Bora Bora without visiting the world-renowned Bloody Mary's Restaurant & Bar. This was our grand finale dinner for the entire trip and a good time we had! Especially the boys, finishing the night downing a couple of bottles of Barossa Valley reds before stumbling down the dock to get into the tenders. Lord only knows what the other patrons were thinking when we attempted to rid Ian of the hiccups. Arms strung up in the air against a pole whilst being fed a glass of water. "Try it with your head upside down and drink backwards, Ian." Howls of laughter, three enormous men crooning island lullabies on their ukuleles under a vast, twinkling thatched-roof restaurant. It was a fantastic, memorable night, though we girls were left wondering if our cocktails actually had alcohol in them. I think NOT. I would not usually be able to stand upright with four piña coladas and a glass of sav blanc under my belt. Glenys most certainly wouldn't. We were suspiciously sober as judges; probably a good thing, though, eh! Nobody makes piña coladas like the Caribbeans!

We ended up leaving Bora Bora a day earlier than planned, caught a stiff breeze, and headed for Niue, smack on the rhumb line. Several boats stayed behind, waiting for the green light from the well-respected New Zealand weatherman, Bob McDavitt. Leaving three days later, they all ran out of wind! The stiff breeze *Marmax* had a fabulous time in was not so great for others. When we got to Niue, we were saddened to hear that a solo-sailing yacht had dismasted behind us while heading out from Bora Bora. Poor guy!

We had a 1048 nautical mile (1900 kilometres) seven to eight-day sail ahead of us to Niue. We would be heading off once we signed out of French Polynesia waters with the gendarmerie. I'll tell you what…when you plot a course for New Zealand from Tahiti, one soon discovers we live a very long way away; we really do live at the bottom of the world!

Bruce had set course for Niue. While he was asleep, Glenys sneakily altered course ever so slightly. We would now run right over the summit of an old volcano that rose four kilometres from the deep oceanic floor. This tiny dot on the chart was called Palmerston Island, a coral atoll in the Cook Islands. At its highest point, it was only four metres high. Glenys and I were extremely keen to explore this little place, as it had a somewhat legendary past thanks to a virile old captain in the mid-1800s. Two hundred yachties a year now called into this fascinating island on their way through to Niue and Tonga; we did not want to miss it! As you can well imagine, Bruce was not too happy when he awoke for his next watch; actually, he was seething. There was hell to pay for that decision.

In our usual form, we arrived at Palmerston at 9 pm at night. We had to kill time plonking about in the moonlight outside the atoll until the following morning. Not knowing the reefs, one obviously did not approach at night. We motored past the Bird Islands then down the reef. Palmerston is the only inhabited island amongst several islands that sit on a continuous necklace of reef protecting the eleven-kilometre-wide lagoon. The island is otherwise known to the villagers as Home Island.

Because of the coral, you cannot get a yacht inside the atoll, so one has to anchor outside the reef. When you arrive, the villagers apparently have a race to your vessel. You are adopted by the family who reaches you first; they then become your hosts. They had seven moorings in ten to fourteen metres of water, which sat one hundred metres off a foaming reef with the surf rolling over it. There was a deep blue drop-off, some seventy to eighty metres deep, just behind our stern. We wondered, did these moorings hold all right? We could see a heavy stainless steel chain secured around a coral bommie through the crystal clear water, then thick rope up to a series of floating buoys. Looked like there would be no problems here.

Andrew arrived to give us a rundown on how things worked around here. Andrew was the son of Bob, whose family were to become our delightful hosts. Customs, immigration, and health officials magically arrived with Bob in his alloy tender through what seemed an invisible pass in the reef. We were thoroughly questioned about our health for the first time on our entire trip. The local nurse was a Papua New Guinean lass. We had to pay $20 for her to do check-ups on us. We now had a collection of money from all around the world aboard *Marmax*, and looking at the nurse's receipt book, we were not alone. She had every denomination you can imagine stapled to the back of each receipt from each visiting boat. Quite a colourful collection there. Clearance completed, we jumped into Bob's boat with the customs party (aka his family), picked up another English couple alongside, and off we headed ashore.

The history of this island was a real eye-opener and a great lesson in the importance of family relationships and genealogy. As this beautiful island story goes, the entire island community was created by William Marsters, a young English adventurer from Leicestershire, UK. He had been working on whaling boats and had become rather sick of life as such and dreamed of living on a deserted South Pacific island. He found himself a Cook Islander wife on Penrhyn Atoll. He then set out to find an uninhabited island of his own, where he could bring up a family and live in peace. He settled on Palmerston Island on July 8th, 1863, with his Polynesian wife and two mistresses, one of which was his wife's cousin. William died at sixty-seven, although his headstone states seventy-eight. He had 23 children and 134 grandchildren!

James Cook discovered and named the island in 1774 after the then British prime minister. Another bloke was floating about when William discovered this lonely island. He had the same idea about claiming the island for himself, but he was a little slow off the mark. When this bloke returned to the island, he discovered William Marsters, fully settled in. With coconut palms, breadfruit, and mahogany tree plantations already planted and a village (split into three for his three ladies and their families) already underway. So he backed off with his tail between his legs, never to be seen again.

William (Papa) Marsters was clearly no dummy. One of the first things that struck us about the people of Palmerston was their intellect. These guys are smart cookies. Entrepreneurial, wicked senses of humour, sharp as tacks. In acknowledging Great Britain had sovereignty over the island, Papa William paid Queen Victoria £50/annum for rent of the entire atoll. Many years later, Queen Elizabeth granted perpetual title to the Marsters family. From there on, they lived happily ever after. Every person living permanently on this island paradise is a Marsters. There are thirty-nine residents at present but has been up to sixty in the past. Many children now leave home for higher education and work in Rarotonga, New Zealand, and Australia.

William created a little village with one extremely wide main street. The remains of the original buildings still stand, including Papa Williams's own home and the church. Papa Williams's house was made of timbers of a wrecked ship and bolted together. It would have been a mean feat simply lugging the wood to the construction site, let alone building it. Much of it would have been done alone, as he had only just started breeding! The three families all have their own separate graveyards. My favourite was the one on the beachfront surrounded by upside-down wine bottles, just like my vegetable garden back home in New Zealand!

After introductions to Bob's family, we were shown around the island by his daughter May. Wandering amongst the groves of coconuts, we came across all kinds of curious things! First past baby frigatebirds perched up in the air for feeding. Old chest freezers EVERYWHERE!!! The islanders' primary form of income was exporting parrotfish. The supply boats only picked them up two to three times a year, so everything was frozen. There was nowhere to dispose of the old, broken-down freezers, so they were stacked all over the show. Most of the freezers were crammed with whole fish, a bit of food, and ice cream. We also discovered loads of frozen Tegel chickens! There were hundreds of chickens running loose around the island. "Why do you not eat those chickens?" we asked.

"No, it's too hard to catch them, too much work," one big dude replied.

We dropped off a big bag of gear for the children we had picked up in Galápagos previously. Pencils, crayons, stickers, balloons, balls, and cool kid's stuff like annoying recorders! The Palmerston Happy School was complete with computers and a satellite link-up for the internet.

There was an impressive solar electricity plant on the island, complements of New Zealand Aid. The half-built Palmerston Yacht Club was built by a gorgeous, spritely old man who introduced himself as Bill Clinton. (Of course, he was a Marsters!) Bill also had a couple of frigatebirds he was feeding. His wife, Metua, cooked us some delicious "KFC" chicken, while Bill animatedly told us about his life on the island. That would be another book alone! What a wonderful, wonderful man. We had an absolute blast with him. He collected flags from all the yachties and swapped them for breadfruit chips, loaves of bread, etc. We'll never forget that guy.

We found the wreck of a yacht called *Ri Ri*. She had been lost on the reef a few years before. Another long story, that one! Anyway, the islanders rescued the owner and tried to save the boat, to no avail. *Ri Ri* of Philadelphia now sits upturned under coconut trees, used as a shelter for filleting fish in the rain. For those nervous about those moorings outside the reef, the islanders' recommendation was to drop a hanging anchor over the bow and let it hang one metre above the seafloor whilst on a mooring. If by chance the mooring did not hold, you then have a second chance at the anchor grabbing something at some time before your hull hit the reef. Good idea 99. *Marmax* duly dropped her anchor over the side for insurance against disaster. Like I said, they are smart.

Bob and his wife cooked us a fantastic island feast. Plates were heaped high with rice, a curried garlic parrotfish cake, lamb chops and johnnycakes, like doughnuts. Absolutely delicious! The family got wind that Les was, as usual, pining for ice cream. Oh boy! I think we got given half a litre of vanilla ice cream each in a bowl. It seemed pretty funny sitting in the middle of absolutely nowhere, on a tropical island, tucking into a giant bowl of NZ's favourite Tip Top ice cream. Bob's youngest two children had skipped home from the Happy School, joined us with the ice cream, but sprinkled half a carton of corn flakes over the top.

After a circumnavigation stroll around the beachline, we reluctantly left the island, and Bob drove us all back through the invisible pass with the surf rolling through it. What fun that return trip was!

The snorkelling from the boat was fantastic, parrotfish and huge almighty coral trout and wrasse all over the place. Visibility underwater was incredible. Les landed a coral trout after he and I were being busted off by these big beasts every time we threw a line in the water. We had given all our fishing line tackle to the islanders, so we were down to our last two hooks when he caught a monster. Heart-racing stuff that put a smile on his face for hours. After being advised by the islanders that there was definitely no ciguatera in the fish here, we decided to fillet this fish for dinner. Delish! We were amazed the islanders ate parrotfish, as they are coral eaters too. We all tucked into the fat coral trout…except Les. He never eats fish the same day he catches it. It's his thing. (Warning! Warning!)

The next day we waited until the wind built up and headed for our very last stop on our ocean adventure, Niue, some 387 nautical miles (717 kilometres) away. Then the trouble started. Glenys went down first. Headaches, joint aches, no energy, no appetite, major in-and-out problems…Ciguatera poisoning from the fish! We all copped it in varying forms. Swollen hands, rashes, sensation opposites, where cold was hot and hot was cold. Totally weird jumping in the water! It stung badly. Six days on, we had fed Glenys all the electrolytes on the boat, and she had almost come right. We were all weak as kittens and full of ibuprofen and antihistamine tablets. The itching was driving us nuts! Consequently, the sail down to Niue was beautiful but full of groaning and lying on the parallel! We were, thank goodness, getting better day by day.

We never expected Niue to be flat for some reason. Sailing from the north, the sun was shining, and she looked absolutely stunning as we approached. A humpback whale waved her tail as we sailed in. Lusciously green, lots of volcanic ledges, sea caves, and a few little sandy beaches sprinkled about. Unfortunately, it was Sunday when we arrived. Customs and immigration were off the air, as with everyone else, such as Niue radio.

You are supposed to get permission to pick up a mooring, as it is extremely deep water, and there are no anchorages, just coral reefs close to the shore. We were out of luck as far as moorings were concerned. All twelve were occupied, so there was no room for us. We tentatively dropped our anchor in the shallowest water we could find and decided to sit it out until the next yacht left a free mooring. Fortunately, a Swedish yacht saw our dilemma, called us on the radio, and informed us they were ready to sail within the next two hours. Anchor up, mooring secured. We were safe for the night. As we could not go ashore without clearance, we all collapsed and rested for the night. Being the action dolls we are, the next day would be a big exploratory one for us. That night, the moored boats all rolled like you would never believe. I could barely cook dinner, stuff flying and crashing everywhere. So sleep was quite out of the question. We were exhausted by morning, super-sensitive to anything going on with this damn ciguatera business; otherwise, I am sure we would have slept like lambs. Such is life.

Early morning found the weather had closed in, and it was raining hard. This was the first time in seven months that we had experienced heavy rain. We knew, now, we were getting closer to New Zealand.

Bruce radioed in at the main town of Alofi, to inform customs and immigration of our desire to get clearance to come ashore. If you had had any form of illness in the past ten days, you had to declare this to authorities. So via VHF radio, Bruce dutifully reported that we had an illness aboard and may need a medical person to attend our clearance. Later, while doing silly things like organising data for our phones, we learned the entire island seemed to have heard this report of an illness aboard the sailing vessel *Marmax*. When we approached her desk, the telecom girl stepped back from us, clearly freaking out that we were contagious. Well…everyone in town knew about us now. Our hire car got dropped off in a car park for us. Funny…they had not asked for driver licenses, ID, nothing. We never ended up ever seeing them! Maybe they too were scared of catching our illness!

Pre-warned about the landing of dinghies, Bruce had rigged our slow-leaking tender up with a three-way sling. This was to deal with Niue wharf. You could secure a crane hook on it and hoist it out of the gnarly water below. First try, we managed it in 1.15 minutes flat. Champions! I made that sound easy, didn't I? Actually, it was rather freaky and not for the faint-hearted. Powerful waves surged in, threatening to slam us into the concrete wall. We had to leap from the wobbling tender, then lunge for the slippery steps, and scamper up to the wharf before the next wave swept over us.

We got customs clearance, no worries. Ciguatera seemed quite common with yachties coming in from Palmerston, so they were very understanding. Apparently, the cargo ship was due in, which was an exciting time for all the islanders. Everyone had two to three jobs on the island, so the customs people had to fit us in with the ships arrival and a plane arriving. In this rain, and the state of our bodies and minds, our first day here was a quiet, "Let's explore the main street and Niue Yacht Club" day.

By the third day, the sun was shining; we were up and at it, all feeling so much better. Off to find the supermarket, the bond store for duty-free (goodbye, another $300), and a tour in our hire van. Wow! Niue's coastline was stunning! Really, truly spectacular!

We walked through the beautiful, lush rainforests of Huvalu, snorkelling in secret sea caves and glowing pools surrounded by luminous pink rocks and the prettiest corals and fish imaginable. We climbed up and crawled down long sea ladders to arrive in a world reminiscent of a *Lost in Space* movie set. We explored mystical pathways to discover deep dark chasms with crystal clear waters over gleaming white sand below. Sighed at the beauty of the tiny, pristine little beaches and the makeshift shacks on the wild oceanfront. We were wowed by the mighty surf breakers smashing onto the huge ancient coral ledges. Fascinated by the beautiful gravesites adorned in tropical flower leis; crumbling old coral brick houses were overgrown with ferns creeping all over them. We skipped amongst floating fairy flowers. (We don't know what they were called, but I bet they inspired the trees in the *Avatar* movie.) This was indeed a heaven on earth.

Catching up with sailing pals at the friendly Niue Yacht Club, we met new ones and feasted like kings. The people there were lovely and welcoming; they giggled a lot. On the very last night, would you believe, we were kept awake by whales singing! As most of us had been suffering aches and pains from the fish poisoning, I woke in the middle of our last night to hear what I thought was Janelle crying. Les, being the healthiest of us, sleepy-eyed, went to look in on Janelle to ensure she was okay. Much to Les's surprise, she was sound asleep. Then what was that emotional sound? Stopping for a toilet break on the way back to his bunk, Les flew into our cabin full of excitement! Here we were, crammed in the dunny with our ears to the hull; the walls had created an auditorium! It was pretty amazing once we realised where the sound was coming from. Niue is famous for whale watching, and our moorings were set right in the middle of the best spot to view them from land.

We spent the last day in Niue having a final exploration. But first, a trip to the hospital to get Glenys checked out and another supply of antihistamine tablets for the rest of us. We walked to the famous Talava Arches. Before this, we simply did not have the stamina to walk what we were told was a two-hour walk through bushes and caves to see this magical place. Ciguatera had a firm grip on us and sapped our usual boisterous vitality! It was a piece of cake. It took us twenty minutes each way, through pristine bush and ocean pathways, carpeted by thick volcanic rocks and clusters of ancient coral. Through breathtaking sandstone caverns, we crawled and popped out on the ocean. An enchanting marine garden opened with the arches in the background, cradled by looming, sheer-faced rock walls.

Glenys, Les, and I, gingerly stepped down over extremely sharp rocks to get to the water. Oh wow! It was like a delicate Chinese garden under the water with incredible sea ferns gently waving in the sea. Rock walls created sea tunnels, tropical fish ebbing in and out, swimming comfortably about our bodies. Perfectly formed corals of pinks, purples, and yellows, living there just as they may have been for hundreds of years, untouched by humans, nothing was broken. It was bewitching. Completely different formations from what we had experienced throughout our journeys of the Caribbean and South Pacific.

As you can see, Niue had really impressed us. Interestingly, she was a strange mix of Galápagos, Marquesas, and New Zealand as far as her topography went. The sharp, volcanic "ouchy-ouchy" rock formations of Galápagos, nowhere near as dramatic in size or height as the Marquesas, but her coastlines were just magnificent. Her ancient rock creations had dramatic energy about them. The Togo Chasm was like a cousin of one of our favourite places on earth, the Baths of Virgin Gorda in the British Virgin Islands. This tiny country was life-be-in-it stuff, and as the natives of Niue said, "This is what it feels like to be connected to the earth, the sea, and the people." There was nowhere like it; we loved Niue!

Come early afternoon, we set a course directly down to New Zealand, hoping to clear in at Whangārei around the 29th of September, weather depending.

Marmax had a bone between her teeth, and she was heading home.

Niue

Moored at Palmerston Island

Leaving Bora Bora

The troublesome fish!

Niue

18. The Home Stretch: Niue to New Zealand

"Bye, Niue!" Gee, it was hard sailing away from the very last country of our fabulous sailing adventure. You know what it's like, saying goodbye to your adult children at international airport gates, not knowing when you will see them again? You walk away with a big lump in your throat, trying to be brave, desperate to hug them just that one more time. It was an emotional day.

Dropping the mooring line in Alofi, Niue, at 4 pm on Thursday the 19th of September, we waved our goodbyes to no one in particular. You always have to wave though, don't you? We sailed off in a perfect breeze, still with our one reef in the mainsail (remember months ago, we blew the main clew out?) and a full genoa.

The first twenty-four hours were brilliant; everything you would dream of for a perfect ride home. A stout 15–20 knots of wind, cruising 7–8 knots through the water, clear starlit skies, a low swell, nice. Day 2 found us with a steady increase in the wind and deteriorating sea conditions, nothing dramatic. At 11.40 pm, Glenys was on night watch again, and we heard a frantic yell for help from the cockpit. Our beautiful, eight-month-old genoa had split her seams; two gaping holes stared back at us in the dark. Great. Though reefed to storm sail size, much to our dismay, a sudden 28-knot gust of wind had managed to beat her up and shred her wardrobe.

Fortunately, Bruce had arranged for a new, all-rounder headsail to be sent from New Zealand to Spain. Our Category 1 rating for the journey required an unlaminated genoa. *Marmax*'s original English sail was laminated. We still had this spare aboard. Thoughts of repairing the new sail were quite out of the question. We were now in a pitching sea and waves foaming over the bow in 20+ knots. In the name of crew (family) safety, we made a call to wait to change sails until the weather settled a little, hopefully in daylight the following morning. That sail was a heavy sucker, and as it was a furling headsail, it had to be unfurled entirely to get it off. The danger, of course, was that one of our precious crew could have a violently flapping clew cause all sorts of damage to their soft head. The thrashing sail could also simply toss any of us over the side. A) None of us was 100% healthy yet. B) The water was now freezing. We were not particularly keen on trying to struggle in the water, getting whacked into a pounding hull whilst strung up like a wet tea bag with our safety harnesses. Fortunately, we had another inner forestay. We were able to raise our bright orange storm jib, a tiny sail, to at least steady the yacht and give her a bit of steerage in the winds.

There is always an "unfortunately" after "fortunately", isn't there? Unfortunately, the weather system held out for another two days, just as we were trying to cross the Tonga Trench just below the Tongatapu Group, part of the Kingdom of Tonga. This was an oceanic trench, the deepest in the southern hemisphere and the second deepest on earth, which formed part of the Tonga-Kermadec Arc, a structural feature of the Pacific seafloor. The Pacific tectonic plate is sneaking below the Indo-Australian plate and lifting it. A fair bit of destructive activity goes on here. Our problem was a roaring tide working against us. There are reported to be around 350 earthquakes here per year, often causing tsunamis and expelling copious amounts of pumice rock. You may have heard of the pumice raft, the size of New York's Manhattan, reported at the time? It was still floating about out there, though we did not see it, as we were probably too far south for this one.

On day 4, morning broke with a significant drop-off in wind strength, down to 7–11 knots. Clear, sunny skies and a gentle ocean swell. The sound of the ocean slipping past the hull sounded like a million kittens lapping up milk out of saucers. Lap, lap, lap, lap! Storm jib off, new genoa removed, and the old, larger spare one put up in its place. We were off again, albeit a bit slow, as we were battling light southerly winds and that surging tide pushing us backwards under a reefed main. We had a couple of bursts of sailing action, but we finally ran out of wind and resorted to steaming with the engine. We could not go all the way to NZ burning diesel, as the last time we had fuelled was in Mo'orea, and without diesel…well, the consequences would be dire. Without fuel, we would have no generator, no power, no autohelm, no charging for our multitude of devices. (C'mon…it is okay to watch the entire series of *Gossip Girls* at sea, isn't it, Janelle?) Anyway, we won't go there. The engine came back on two hours later; it was turned off to try and catch a light breeze. On, off. On, off…Grrrrrr!

Marmax may have headed off home with a bone between her teeth. Still, like all big dogs, she appeared trapped at a fence, running up and down, trying to get through into the paddock where New Zealand was hiding out! Where the heck had the wind gone? Sails flapping listlessly, gasping for a breath of air. You know that annoying squeak in your car? It gets magnified down a nineteen-metre aluminium mast. The silence of floating about on a flat, glassy ocean with an occasional swell would drive you nuts if you could not disguise it. Fortunately, we are a noisy family, but sometimes, especially at night, the monotony of the sounds and the sheer boredom of no action does get on your goat.

One thing that did keep us going was keeping up the heart pump for the generator; you know that chicken-baster type balloon you have to squeeze to get fuel moving through the lines? Yes, we were still under jury rig, manually pumping fuel from the main tanks into a tote tank. Why a tote tank? So it could be put under the galley stairs and gravity fed into the generator to keep everything going instead of hand pumping it. Annoying in bouncy weather, to say the least, but it worked, so we were grateful for that! Every two hours, twenty-four hours a day, for weeks.

Ants-in-your-pants Glenys was constantly moving about, cleaning, chattering, writing endless emails. She is like a wind-up Eveready battery bunny until her key stops winding and she collapses for a thirty-minute kip. The ciguatera business was now over apart from wild itchiness! I had been on antihistamines for over a month now. *Marmax* College was into its final days of revision and practice before the Year 10 exams back in New Zealand. We were still sticking to a regular five-hour day regime. Janelle was now scoring A's and B's from Les and me, a far stretch from the beginning of the year! There was certainly something to be said about home-schooling for some kids.

While we struggled through this weather fence, anyone watching our PredictWind tracker would wonder if we were sober drivers. We took turns drawing shapes with our chaotic course, thanks to the fickle winds and tides. Bruce found a way of doing a 360 while doing 1.7 knots of speed; Les did a couple of 180s. Boredom was setting in! There was a debate going on. I thought we had made a map of Queensland, Janelle thought it was a fish, and Glenys, a pig with boobs.

The kittens were batting us like a ball of wool. We needed to be sailing directly to New Zealand to make up lost time and quickly. It was frustrating, especially for Glenys.

Bruce, Glenys, and Janelle had a dear friend who had died just before we reached Niue. His funeral was delayed until the 4th of October. Our ETA was to arrive in Whangārei before 1st October. Initially, we thought this would be no problem. Our hopes of reaching New Zealand in time to attend the funeral were quickly slipping away. Glenys and Janelle wrote their memoirs in preparation.

With night now falling upon the sea, the sun scrabbled for a finger-hold on the edge of the horizon. Blue, bruised purple, and brilliant hues of oranges and lemons…another silent night punctuated by the sound of those kittens licking their plates.

"Creak, creak." Come on, wind! Get your act together! I came on watch at midnight, my usual time slot. The golden moon rose in the distance; a big grin on his face flicked on a million lights above. We would have to be content stargazing that night to keep ourselves awake. These were to be the last stars we saw on the trip.

Les woke me up to excitedly tell me about a huge, white moonbow that had mysteriously appeared on his watch. My eldest son trained as a pilot and later told us its name, as we had never heard of them. They are apparently a rare sight and are also called lunar rainbows. They are caused when the light is reflected off the moon onto water droplets in the air. The brighter the moon, the brighter the moonbow.

They are so rare because, usually, the moonlight is not bright enough to reflect the water and, quote, "excite the colour receptors in human eyes" like a rainbow can. It was pretty cool!

Two enormous albatrosses joined us; what a marvellous sight they were. It was getting colder and colder as we slowly slipped closer to Antarctica. Out with the beanies, jerseys, and wet-weather gear; oh well…Les would be happy. An eight-hour stint on the engine, charging through the water. There was a realisation that we were definitely not going to be able to make the funeral. Speeches were emailed to the family to speak on behalf of Glenys and Janelle on the day. New Zealand customs and immigration had been notified of our four- to five-day delay. We had to give forty-eight hours' notice to them before arriving; otherwise, we could collect a $400 fine. A nice way to welcome native seafarers home, that's for sure. We guessed they must send up spy planes or something once you reach Cape Reinga to make sure you pass go and land in the right hands; perhaps we could stop off for a BBQ on the beach on the way down?

Finally, away from the effects of the Tonga Trench, we were running with a southerly current. Hallelujah! The wind was right up our stern, but we were moving! We poled out the genoa; it was a beautiful sail.

We celebrated by making Kiwi dip to accompany our final ration of Sunday potato chips. One thing we waited seven whole months for was a can of Nestle reduced cream! Glenys and I looked high and low, through every supermarket, in every country visited, for this stuff. How good is Maggi Kiwi dip, folks? Coupled with a bottle of NZ sav blanc and I was in heaven! The last of the Hinano beer and Tahitian rum was finally coming to a delicious end.

By now, we were officially over rice and pasta. One thing Niue did not have was fresh vegetables. Captain Bruce, impatient to get home to New Zealand, would not allow us to stick around for the Niue supply ship to be unloaded. Consequently, we were now down to half a solitary onion. I believe I used onions in every single dinner I made aboard *Marmax*! We had one very miserable-looking potato. This one only survived because it jumped out of the vegetable nets under the cockpit steps sometime back in Bora Bora days. It hid under the emergency grab bag – under the navigation table – like a mouse hiding out. We also had one shrivelled French carrot and a lonely lime from up the river in Raiatea. Strewth! We were on rations now!

The cooking gas ran out around this time too. Thank goodness we had purchased an electric paella pan in the BVIs. With two centimetres of depth in the pan, you can only imagine how hard it was to cook in this thing in a raging sea on top of a gimbal stovetop. One night, I shallow fried battered tinned corned beef for dinner on top of herby rice. Don't ever tell Janelle. She had nightmares about Spam and believed that canned corned beef was a direct relative. It was nothing a good daub of tomato or chilli sauce could not fix, especially if it was served in the cockpit, in the dark. Glenys fed the crew a hearty Christmas pudding with custard at 4 pm snacks; yes, she did! That would take the edge off our hunger!

We woke up to our eighth punch-up on board. "A pinch and a punch for the first of the month, no returns!" hollered Janelle, as she pummelled us unawares. There was much screaming and retaliation going on. Sometimes all you could do was cower and kick yourself silently inside that you weren't the first to think of it. It was the start of the eighth month at sea for Les and me and the ninth for Glenys, Bruce, and Janelle.

The wind picked up, pushing us to a speed of 7–8 knots. Snap, crackle, pop! *Marmax* was off and racing with her teeth back and bloodshot eyeballs pointing towards New Zealand; it was exhilarating!

We anxiously awaited the predicted onslaught ahead. New Zealand was surrounded by bright red storm patterns. In the pit of our stomachs, we knew we were about to cop it…

The seas start flattening in a ripping wind, building up to just over 40 knots. We were fully reefed with three down on the main and a storm jib on. What happened to our friendly moon? Temperatures plunged. Glenys and I counted eighteen pieces of clothing being worn between us. Sea boots were on, plus gloves. A four-hour stint on watch between midnight and 4 am found me without feeling in my hands or feet. Try stepping down steps backwards in a boat while being flung from side to side on a surfboard and you can't feel where your feet are! It was getting too dangerous for one-man watches, so we doubled up. We were sleeping wherever we could go in full wet-weather gear. The terms cold, wet, and hungry took on a new meaning. *Marmax*'s bow began free-falling over the crests of swells. You could feel them coming as your whole body rose up in the air. It was the crash at the bottom that jarred your teeth and then the release of another sigh of relief with the realisation that nothing on the boat was breaking, cracking, or falling apart. The mast was now squirting water out of the collar in the saloon, water sliding across the interior bulkheads, making the floor dangerously slippery. Now we had not only wet clothes everywhere, but the upholstery was soaking. *Marmax* handled the sea well. At around 21 tonnes loaded, she sliced through most of the chaos; she had never been tested like this before.

Marmax's deck windows now had green water flowing past. I often wondered how Janelle would cope with a quiet classroom when she got back to college the next term, just sitting in a dumb chair. Foaming water was streaming past the window behind her head; we braced ourselves for every slam into the water, pencil cases were zipped up, ready to fly. We had our hands clamped over heavy schoolbooks and, of course, our precious dictionary! Knees up on the table and jammed in with them around our chins. It was no fun cooking in that electric paella pan. So, folks…rice or pasta? Or would that be pasta or rice? God, I missed onions! We could not catch fish either. Niue Hospital had told us not to eat any fish for…how long? Six months? You are kidding me!

The wind changed its mind. It was blowing straight out of the Bay of Islands, now direct from our destination, Whangārei. What a slog! Tide, massive waves, and screaming winds were pushing us backwards on a starboard tack. We were forced to head downwards towards Great Barrier Island on a port tack. Friends and family back home watching our track must have wondered what we were up to. We heard there was much speculation about our reasoning to head so far south; we could not blame them, as even we were in a state of perplexity!

Dozens of bioluminescent glowing sea worms/eels/squid, or whatever the heck they were, floated just below the surface, often well above the cockpit as the swells were building. Some were up to a metre long like fluorescent light tubes floating past; they threw out so much light! Les had come out on watch and had nearly jumped out of his skin. In his half-awake state, he thought we had just run over a submarine. Pitch-black, the phosphorescence of our wake was mesmerising despite the unfriendly weather. Next time we do an overseas yacht passage, I swear I will take along a fancy camera that will take photos of the strange world of ocean nightlife. For the entire trip, it had been incredible.

Where was the Land of the Long White Cloud? We should have had a visual on her by now, but we were surrounded by angry, low clouds. The GPS and radar would have to guide us home.

We were now 12 nautical miles off Great Barrier Island, 40 nautical miles from Whangārei. It was time for the final tack home. The wind was blowing 30 knots straight at us; we could not lay the line, and for the first time on our long journey, we all agreed that we were over this atrocious weather welcome. With a furled headsail, the iron horse got a kick in the guts, and we motored the last few hours home as the day was dawning over Whangārei Heads in the distance. One last backwards gaze at the wake of *Marmax*'s stern flowing along the deep blue sea…memory bubbles leaving a glorious trail behind us…

Then all was calm; we were embraced safely into the quiet of the harbour. It was dead still, eerily still, and freezing, icy cold. The mighty Mount Lion, Mount Manaia, and Mount Aubrey saluted us as we dropped the mainsail and glided past Marsden Point. We silently slid alongside the quarantine berth in the early morning light. We had made it! How good it was to be safely home.

It had taken us fourteen days to get from Niue to Whangārei, New Zealand, way too long! The entire trip from Spain to New Zealand had taken 218 days, covering 23,053 nautical miles, which is 46,648 kilometres. We had arrived feeling fit as fiddles, tanned and exhilarated, and on average, we each lost around ten kilograms of body weight…Needless to say, I have since found my ten again!

Customs and immigration officials arrived right on time. Biosecurity gave us a thorough search through all of our supplies and gear, then we were cleared. (Goodbye, Tuamotu brain coral… sniff-sniff!) Because *Marmax* was classified as an imported vessel from Spain, a substantial GST and import duty was payable. Being a Friday, there was no way this could be sorted by either the banks or the New Zealand Customs Service. We were locked securely under customs bond, behind tall security gates, away from our support team, who had started arriving. We were also securely locked out from the boat. Some swift negotiating with the friendly customs guy found us permitted to jump two berths to the other side of that big quarantine gate.

"Hellooooooooo…!!!!!" That voice was familiar! Our younger sister surprised us and had arrived with Mum. They could only get as close as the other side of the waterway after breaking into the lawns of someone's private waterfront property. We yelled excitedly at each other across the way, much to Bruce's obvious annoyance and embarrassment. I know. Us girls are loud! Jenny and John, Craig and Carey, such a joyous welcome with lots of fresh food! We popped the champagne, never stopped talking for a few hours, and had a merry time! We were to go to dinner at the marina restaurant; somehow, plans changed. The next thing we knew, we were hastily packing a few essentials we had aboard, got bundled into cars, and set off home. Glenys and I whispered our words of thanks to *Marmax* before leaving her alone in the marina. Yes, like many of you, Glenys and I firmly believe boats have feelings. She had gotten us home safely, and we will remain ever grateful to have been blessed with luck and laughter all the way.

Heading Home...Goodbye Niue!

Home Sweet Home, Whangarei Heads, New Zealand

We made it!

Home

How good was it to see the green, green grass of home! Newborn lambs sleepily watched us come up the driveway of the farm. There were birds everywhere, an explosion of spring colours, reunited with our much-loved dogs...It was a strange, fabulous feeling being home.

Five months after arriving in New Zealand, our lives, like yours, were interrupted by the chaos of COVID-19 and, more recently, the effects of climate change. We will forever be grateful to not only have arrived safely home, but we completed our grand voyage before the world started shutting down. We often think of all those courageous sailors we met along the way who had sold up their homes and run off to sea for a lifetime of adventure. Are they still being welcomed with those friendly smiles? We think not. It is gratifying to know we were one of the very last sailing families to have made it home safely before our perfect planet collapsed. The world appears to have changed forever.

No matter what, the surreal afterglow of this epic family sailing adventure will never, ever leave us. We thank our lucky stars every single day.

As the famous Walt Disney once said, "If you can dream it, you can do it…"

Cheers from the *Marmax* crew!
Bruce, Glenys, Janelle, Les, and Debbie

APPENDICE

MARMAX SPECS

Oyster 46 Sailing Cruiser

Designed by Rob Humphries

Built 2008

Specifications:

Length Overall (including pulpit): 14.26m - (46' 10")

Length of hull: 13.94m - (45' 9")

Length of waterline: 12.36m - (40' 7")

Beam: 4.41m - (14' 6")

Draft HPB keel: 2.16m - (7'1")

Rig and spar type: Masthead sloop with fully battened mainsail

Keel: High Performance Bulb (HPB) external lead keel

Rudder: Fully protected skeg-hung

Displacement: 17.5 tonnes (38,581 lbs)

Tanks - fuel: 750 litres, (165 Imp gals, 198 US gals)

Tanks - water: 650 litres, (143 Imp gals, 172 US gals)

Air Draft (excluding Antenna) 19.03m

Engineering

Engine: Volvo D2-75, 4-cylinder 55kW (75Hp), turbocharged diesel engine with shaft drive

Prop: Bruntons 20" 3-bladed Varifold folding propeller.

Rope Cutter: Ambassador rope cutter on shaft.

Alternator: 14V/115A alternator.

Generator: Mastervolt WhisperGen M-SC6 generator

Bow Thruster: Max Power CT 100 (7Hp) tunnel bow thruster with eyebrow features.

Steering system: A Whitlock cable unit installed with a remote Constellation 'Royale' pedestal.

Air Conditioning: Clima Marine Vega MkII reverse cycle in saloon and owners cabin.

Heating: Webasto Thermo 90 warm air heating with 5 fan coil units: two in saloon and one in each cabin

Shore power inlet, Generator, Inverter, water maker

Tankage

FUEL: 750 litres approximately (165 Imp gallons/198 US gallons) in a single GRP tank.

WATER: 650 litres approximately (143 Imp gallons/172 US gallons) in a single GRP tank.

TANK TENDER GAUGES: Fuel and water tank tender gauges.

HOT WATER: From an insulated, approximately 50 litres (11 Imp gallons/13 US gallons) tank

heated by the engines heat exchanger or by thermostatically controlled 220V,

1.25kW immersion heater elements powered by the generator or dockside power

Pressurised hot and cold freshwater system

120 litre an hour 240-volt water maker – Village Marine

HOLDING TANKS: Polyethylene gravity holding tanks for forward and aft heads with Wema gauge and deck pump out.

GREY WATER: Each head compartment is fitted with an electric Whale Gulper pump, operated by a waterproof rocker switch. This evacuates water from the washbasin and shower directly overboard.

The galley sinks discharge with an electric Whale Gulper pump, operated by a waterproof rocker switch.

Construction

Hand lay-up GRP hull and deck - Outer hull laminate in Vinylester resin

Hull style lines - 1 x cove and 2 x boot top lines

Anti-foul – Copper coated

High-Performance Bulb external lead keel

Fully protected skeg-hung rudder

EURCD Design Category 'A' Ocean

Oyster Deck Saloon with single centre cockpit layout and single wheel steering

Cored deck and coach roof laid with teak

Spacious, self-draining teak-laid cockpit, comfortable and protected

Deck Layout and Equipment

Anchor: Rocna 33kg. Fortress FX23 kedge anchor

Windlass: Lewmar V4 12V vertical type electric windlass. Deck mounted foot switches and remote windlass controls in the cockpit.

Deck Access: Midship boarding gateways

Stainless steel pulpit

Gate to pushpit

'Sugar scoop' stern with teak-laid bathing platform with hot and cold deck shower and stainless-steel transom ladder

Hatches: Vertical, sliding acrylic washboard for the main

companionway hatch

Lewmar deck hatches 4

Opening hull and coach roof ports 7
Oyster deck saloon windows - 2 forward windows open with gas strut supports

Lewmar opening port into the cockpit (galley side).

General:

Hull to deck joint with teak cap rail

Stainless steel mooring cleats 6

Sprayhood with stainless steel frame over the main companionway

Bimini

Stainless steel stern pushpit seats

Stainless steel life raft stowage

Stainless steel grabrails atop deck saloon

Lewmar deck gear, including racing blocks

Stainless Steel grab handle to aft hoop of sprayhood frame with leather cover

Large lazarette in the stern deck with teak laid GRP hatch

Gas locker for two bottles

Whitlock pedestal with Autohelm, cable steering and SIRS compass

Access to sail handling controls from the cockpit

Davits

Rig & Sail Handling

MAST and RIGGING:

Sloop rig

Seldén silver anodised aluminium spars with slab reefing

Twin spreader masthead rig

Solid rod kicker

Harken B502 backstay tensioner

Harken track

Removable inner forestay

Standing rigging stainless steel 1 x 19 wire

Mainsail reefing lines rooted back to the cockpit (three reefs)

HEADSAIL FURLING: Harken Mk III manual furling gear

WINCHES By Lewmar:

2 x 58 CEST electric two-speed self-tailing primary winches

2 x 46 CST self-tailing secondaries

1 x 46 CEST electric two-speed mainsheet

1 x 46 CST two-speed main halyard on the mast

1 x 46 CST two-speed genoa halyard on the mast

1 x 46 CEST electric two-speed reef line in cockpit.

SPINNAKER GEAR: Seldén single spinnaker pole stowed vertically on the mast

SAILS:

Fully battened mainsail including lazy jacks

Furling genoa

Cruising chute

Top-down furler and removable carbon bowsprit

Storm jib

New Genoa # 2

All standing and running rigging replaced in Barcelona

Navigation & Communication

COMPASSES: SIRS Major 150mm binnacle-mounted steering compass

SAILING: Raymarine ST60+ sailing instrument pack – MFD, speed, analogue wind and depth

INSTRUMENTS: Displayed in GRP cockpit console + Raymarine ST60 and Graphic display at chart table

CHART PLOTTER/RADAR/GPS: Raymarine E120 chart plotter/radar/GPS at chart table, connected to 4kW radome on the mast

Raymarine C97 digital chart plotter

Raymarine Raystar 125 antenna

Waterproof hard case for chart plotter mounted at the helm

AIS: Raymarine AIS 500 transponder

AUTOPILOT: Raymarine ST7000 autopilot with ST6001+ control at the pedestal

WINDEX: Windex at the masthead with combined VHF aerial.

NAVTEX: Furuno Navtex Plus, including GPS interface and aerial on pushpit

VHF RADIO: Raymarine 240E Class D-DSC VHF radio/telephone with masthead antenna and repeater speaker in the cockpit

Icom IC-M33 handheld VHF

ACCOMMODATION

6 berths in a 3-cabin configuration + 2 sea berths in the saloon

Interior:

Forward Cabin:

V-berth with a filler piece. Storage in lockers and drawers. Split mattress with lee cloth. Large Lewmar deck hatch with OceanAir sky screen for ventilation and privacy

Twin Cabin to Port:

Twin-bunk berths with storage beneath and lee cloths. Hanging locker. Lewmar hatch with OceanAir sky screen

Guest Head to Starboard:

Manual Jabsco head. Vanity with Avonite countertop. Mirror and storage. Receiver-style shower and curtain

Owners Cabin:

Island berth with drawer storage underneath. Hanging locker, cabinets and desk/vanity table

Custom composite foam mattress to owner's aft double berth

Opening hull port windows. Large Lewmar overhead hatch

Ensuite head with shower stall. Jabsco toilet with holding tank

Main Saloon:

Oyster deck saloon with seating and dining area utilising the entire width of the hull

Large L-shaped settee to port with a white oak dining table that folds across the saloon to accommodate dining for six or seven. Bottle stowage locker beneath the table. All settee bottoms and backs are removable to access storage areas and systems. To starboard is a large base cabinet locker forward of the settee. Bar locker with glass storage. Trademark Oyster opening forward saloon windows with OceanAir blinds

Navigation Station:

Forward-facing chart table with opening top. Instruments are mounted in removable panels

Electrical distribution panel for DC and AC loads on the bulkhead behind navigation station.

Galley:

Walk-through galley to port with Force 10 4-burner propane stove-top and oven. Microwave. Twin stainless-steel sinks. Refrigerator/freezer

Opening Gebo port to the cockpit, opening Gebo hull port outboard of the stove

White Oak joinery

Shadow Gap joinery style

Flush/concealed hinges on locker doors

Grain matched locker doors and joinery fronts

Oyster custom locker door fittings

Galley worktops in Avonite finish

WC countertops in Avonite finish

Teak sole boards, rubber-mounted for sound insulation

Lee cloths on all sea berths

Alcantara upholstery in saloon

Oceanair blinds to all overhead accommodation deck hatches

Oceanair blinds to all deck saloon windows

Airconditioning

Heller fans

All light bulbs changed to LEDs

Safety Gear

Offshore Medical Kit - Cat 1

Various additional First aid kits

Emergency Grab Bag

Flares, including electronic LED Red flare

Raymarine Man overboard (MOB) Life tag system with wrist tags

Raymarine MOB button in the cockpit

Life rings x 2

Man overboard pole

Life rafts x 2

Personal Locator Beacons

Jack lines and plenty of cockpit strongpoints

Large emergency sea anchor

large fenders x 9

Inflatable tender + outboard engine

Inflatable kayak

Radar

AIS-Automatic Identification System

Navigation plotters. Two hardwired plus on three phones, 2 laptops and 3 tablets

Emergency Position Indicating Radio Beacon (EPIRB)

Dan-Buoy: for the purpose of marking a yacht's/persons location

Jonbouy recovery module

Mast mounted radar and VHF antennas

Emergency tiller steering

Navigation lights (Deck mounted and masthead mounted) + backups

Sheathed knives strategically placed

Self-draining cockpit

Compasses – Fixed and handheld

Personal Floatation Devices (275N) with built in Safety harness

Sea boots and gloves

VHF Marine radio + handhelds

Iridium satelite phones

Mobile phones

Manual bailing system

Bilge pumps

Raymarine Autohelm

Distress sheets

Floating rescue ropes

Air horns – gas canister and manual

Searchlights + torches & head torches

Lee cloths

Emergency hatches/ fire escapes

Sunglasses + safety glasses

Bolt cutters

Battery angle grinder

Fire blankets

Fire extinguishers

Spare engine parts

Storm sail

Sail repair kit

Charts

Duct tape

Fuel and battery shutoff switches

Acknowledgements

We could never thank every person who followed us personally and helped in every little way to get us through this adventure.

Thank you to those who followed us on Facebook during our journey. To the young and not so young people who are so busy in their own lives, we hope your quick reads and the vlogs inspired you to do something unique and gutsy in your lives.

Thank you, Craig and Carey, for anchoring up at Great Barrier that momentous day so long ago, for your friendship and for always being there at the end of an email for Bruce. Brett Whitmore for enthusiastically following us and logging our travels depicted from PredictWind and our Facebook posts. To our many friends and beautiful family. I simply cannot imagine what it was like for our kids in Barcelona, waving goodbye to us as we left on a lonely little yacht sailing off to New Zealand!

A big shout-out to our little sis, Helen, and hubby, Jim, for holding the fort at home on the farm for Mum and watching over our yacht, *Shadow*. Bruce Thompson for caring for Bruce and Glenys's other yacht, *Arctic Lady*. A big thanks to Andrew Dobson for also taking care of Tongue Farm and Woody in our absence. And to Joy for minding Glenys and Bruce's home with their dog, Tara. Thank you for all the beautiful messages we received along the way from our followers.

Jenny and John, for being part of the homecoming team! Donna Wyllie for watching over my clients. Members of the Sandspit Yacht Club and the Algies Bay Boating Club, many of whom have always dreamed of making such a journey, thank you so very much. Thanks to our Mum for believing in us and supporting our folly. She is an incredible woman. Without all of this help and support, the entire trip would simply not have been possible.

Can You Help?

Thank You for Reading My Book!
I really appreciate all of your feedback,
and I love hearing what you have to say.
Please leave me an honest review on Amazon
letting me know what you thought of it.

Thanks so much!

Debbie